THE APPLIED TENSORFLOW AND KERAS WORKSHOP

WORKSHOP

Develop your practical skills by working through a real-world project and build your own Bitcoin price prediction tracker

Harveen Singh Chadha and Luis Capelo

THE APPLIED TENSORFLOW AND KERAS WORKSHOP

Authors: Harveen Singh Chadha and Luis Capelo

Reviewers: Abhranshu Bagchi, Achint Chaudhary, Vishal Chauhan, Alexis Rutherford, and Subhash Sundaravadivelu

Managing Editor: Anushree Arun Tendulkar

Acquisitions Editors: Sneha Shinde and Karan Wadekar

Production Editor: Roshan Kawale

Editorial Board: Megan Carlisle, Samuel Christa, Mahesh Dhyani, Heather Gopsill, Manasa Kumar, Alex Mazonowicz, Monesh Mirpuri, Bridget Neale, Dominic Pereira, Shiny Poojary, Abhishek Rane, Brendan Rodrigues, Erol Staveley, Ankita Thakur, Nitesh Thakur, and Jonathan Wray

First published: July 2020

Production reference: 2240221

ISBN: 978-1-80020-121-7

Published by Packt Publishing Ltd.

Livery Place, 35 Livery Street

Birmingham B3 2PB, UK

WHY LEARN WITH A PACKT WORKSHOP?

LEARN BY DOING

Packt Workshops are built around the idea that the best way to learn something new is by getting hands-on experience. We know that learning a language or technology isn't just an academic pursuit. It's a journey towards the effective use of a new tool—whether that's to kickstart your career, automate repetitive tasks, or just build some cool stuff.

That's why Workshops are designed to get you writing code from the very beginning. You'll start fairly small—learning how to implement some basic functionality—but once you've completed that, you'll have the confidence and understanding to move onto something slightly more advanced.

As you work through each chapter, you'll build your understanding in a coherent, logical way, adding new skills to your toolkit and working on increasingly complex and challenging problems.

CONTEXT IS KEY

All new concepts are introduced in the context of realistic use-cases, and then demonstrated practically with guided exercises. At the end of each chapter, you'll find an activity that challenges you to draw together what you've learned and apply your new skills to solve a problem or build something new.

We believe this is the most effective way of building your understanding and confidence. Experiencing real applications of the code will help you get used to the syntax and see how the tools and techniques are applied in real projects.

BUILD REAL-WORLD UNDERSTANDING

Of course, you do need some theory. But unlike many tutorials, which force you to wade through pages and pages of dry technical explanations and assume too much prior knowledge, Workshops only tell you what you actually need to know to be able to get started making things. Explanations are clear, simple, and to-the-point. So you don't need to worry about how everything works under the hood; you can just get on and use it.

Written by industry professionals, you'll see how concepts are relevant to real-world work, helping to get you beyond "Hello, world!" and build relevant, productive skills. Whether you're studying web development, data science, or a core programming language, you'll start to think like a problem solver and build your understanding and confidence through contextual, targeted practice.

ENJOY THE JOURNEY

Learning something new is a journey from where you are now to where you want to be, and this Workshop is just a vehicle to get you there. We hope that you find it to be a productive and enjoyable learning experience.

Packt has a wide range of different Workshops available, covering the following topic areas:

- Programming languages
- Web development
- Data science, machine learning, and artificial intelligence
- Containers

Once you've worked your way through this Workshop, why not continue your journey with another? You can find the full range online at http://packt.live/2MNkuyl.

If you could leave us a review while you're there, that would be great. We value all feedback. It helps us to continually improve and make better books for our readers, and also helps prospective customers make an informed decision about their purchase.

Thank you,
The Packt Workshop Team

Table of Contents

Chapter 2: Real-World Deep Learning: Predicting the Price of Bitcoin   33

Chapter 3: Real-World Deep Learning: Evaluating the Bitcoin Model   71

Chapter 4: Productization   107

PREFACE

ABOUT THE BOOK

Machine learning gives computers the ability to learn like humans. It is becoming increasingly transformational to businesses in many forms, and a key skill to learn to prepare for the future digital economy.

As a beginner, you'll unlock a world of opportunities by learning the techniques you need to contribute to the domains of machine learning, deep learning, and modern data analysis using the latest cutting-edge tools.

The Applied TensorFlow and Keras Workshop begins by showing you how neural networks work. After you've understood the basics, you will train a few networks by altering their hyperparameters. To build on your skills, you'll learn how to select the most appropriate model to solve the problem in hand. While tackling advanced concepts, you'll discover how to assemble a deep learning system by bringing together all the essential elements necessary for building a basic deep learning system - data, model, and prediction. Finally, you'll explore ways to evaluate the performance of your model, and improve it using techniques such as model evaluation and hyperparameter optimization.

By the end of this book, you'll have learned how to build a Bitcoin app that predicts future prices, and be able to build your own models for other projects.

AUDIENCE

If you are a data scientist or a machine learning and deep learning enthusiast, who is looking to design, train, and deploy TensorFlow and Keras models into real-world applications, then this workshop is for you. Knowledge of computer science and machine learning concepts and experience in analyzing data will help you to understand the topics explained in this book with ease.

ABOUT THE CHAPTERS

Chapter 1, Introduction to Neural Networks and Deep Learning, gets us to select a TensorFlow-trained neural network using TensorBoard and trains the neural network by varying epochs and learning rates. This will give you hands-on experience of how to train a highly performant neural network and allow you to explore some of its limitations.

Chapter 2, Real-World Deep Learning: Predicting the Price of Bitcoin, teaches us how to assemble a complete deep learning system, from data input to prediction. The model created will serve as a benchmark from which we will be able to make improvements.

Chapter 3, Real-World Deep Learning: Evaluating the Bitcoin Model, focuses on how to evaluate a neural network model. Using hyperparameter tuning, we will improve the performance of the network. However, before altering any parameters, we need to measure how the model performs. By the end of this chapter, you will be able to evaluate a model using different functions and techniques.

Chapter 4, Productization, covers the handling of new data. We will create a model that is able to learn continuously from the patterns it's shown and make better predictions. We will use a web application as an example to demonstrate how to deploy deep learning models.

CONVENTIONS

Code words in text, database table names, folder names, filenames, file extensions, pathnames, dummy URLs, user input, and Twitter handles are shown as follows:

"After activating your virtual environment, make sure that the right components are installed by executing **pip** over the **requirements.txt** file."

Words that you see on screen (for example, in menus or dialog boxes) appear in the same format.

A block of code is set as follows:

```
$ python -m venv venv
$ venv/bin/activate
```

New terms and important words are shown like this:

"In machine learning, it is common to define two distinct terms: **parameter** and **hyperparameter**."

CODE PRESENTATION

Lines of code that span multiple lines are split using a backslash (\ ). When the code is executed, Python will ignore the backslash, and treat the code on the next line as a direct continuation of the current line.

For example:

```
history = model.fit(X, y, epochs=100, batch_size=5, verbose=1, \
                    validation_split=0.2, shuffle=False)
```

Comments are added into code to help explain specific bits of logic. Single-line comments are denoted using the # symbol, as follows:

```
# Print the sizes of the dataset
print("Number of Examples in the Dataset = ", X.shape[0])
print("Number of Features for each example = ", X.shape[1])
```

Multi-line comments are enclosed by triple quotes, as shown below:

```
"""
Define a seed for the random number generator to ensure the
result will be reproducible
"""
seed = 1
np.random.seed(seed)
random.set_seed(seed)
```

SETTING UP YOUR ENVIRONMENT

Before we explore the book in detail, we need to set up specific software and tools. In the following section, we shall see how to do that.

INSTALLATION

The following section will help you to install Python in Windows, macOS, and Linux systems.

INSTALLING PYTHON ON WINDOWS

Python is installed on Windows as follows:

1. Ensure that you select Python 3.7 (for compatibility with TensorFlow 2.0) from the download page on the official installation page at https://www.anaconda.com/distribution/#windows.

2. Ensure that you install the correct architecture for your computer system; that is, either 32-bit or 64-bit. You can find out this information in the **System Properties** window of your OS.

3. After you download the installer, simply double-click on the file and follow the user-friendly prompts on screen.

INSTALLING PYTHON ON LINUX

To install Python on Linux, you have a couple of good options; namely, Command Prompt and Anaconda.

Use Command Prompt as follows:

1. Open Command Prompt and verify that **p\Python 3** is not already installed by running **python3 --version**.

2. To install Python 3, run this command:

```
sudo apt-get update
sudo apt-get install python3.7
```

3. If you encounter problems, there are numerous resources online that can help you to troubleshoot the issue.

Alternatively, you can install Anaconda Linux by downloading the installer from https://www.anaconda.com/distribution/#linux and following the instructions.

INSTALLING PYTHON ON MACOS

Similar to Linux, you have a couple of methods for installing Python on a Mac. One option for installing Python on macOS X is as follows:

1. Open the Terminal for Mac by pressing *CMD* + *Spacebar*, type **terminal** in the open search box, and hit *Enter*.

2. Install Xcode through the command line by running **xcode-select --install**.

3. The easiest way to install Python 3 is by using Homebrew, which is installed through the command line by running **ruby -e "$(curl -fsSL https://raw.githubusercontent.com/Homebrew/install/master/install)"**.

4. Add Homebrew to your **$PATH** environment variable. Open your profile in the command line by running **sudo nano ~/.profile** and inserting **export PATH="/usr/local/opt/python/libexec/bin:$PATH"** at the bottom.

5. The final step is to install Python. In the command line, run **brew install python**.

Again, you can also install Python via the Anaconda installer, available from https://www.anaconda.com/distribution/#macos.

INSTALLING PIP

Installing Python via the Anaconda installer comes with **pip** (the package manager for Python) pre-installed. However, if you have installed Python directly, you will have to install **pip** manually. The steps to install **pip** are as follows:

1. Go to https://bootstrap.pypa.io/get-pip.py and save the file as **get-pip.py**.

2. Go to the folder where you have saved **get-pip.py**. Open the command line in that folder (that's Bash for Linux users and Terminal for Mac users).

3. Execute the following command in the command line:

```
python get-pip.py
```

4. Please note that you should have Python installed before executing this command.

5. Once **pip** is installed, you can install the desired libraries. To install pandas, you can simply execute **pip install pandas**. To install a specific version of a library, for example, version 0.24.2 of pandas, you can execute **pip install pandas=0.24.2**.

JUPYTER NOTEBOOK

If you have not installed Python via the Anaconda installer, you will need to install Jupyter manually. Refer to the *Alternative for experienced Python users: Installing Jupyter with pip, section* at https://jupyter.readthedocs.io/en/latest/install.html#id4.

JUPYTERLAB

The Anaconda distribution includes JupyterLab, which allows you to run Jupyter Notebooks. Jupyter Notebooks are accessed via your browser and allow you to interactively run code as well as embed images and text in an integrated environment.

INSTALLING LIBRARIES

pip comes pre-installed with Anaconda. Once Anaconda is installed on your machine, all the required libraries can be installed using **pip**, for example, **pip install numpy**. Alternatively, you can install all the required libraries using **pip install -r requirements.txt**. You can find the **requirements.txt** file at https://packt.live/3haRJp0.

The exercises and activities will be executed in Jupyter Notebooks. Jupyter is a Python library and can be installed in the same way as the other Python libraries – that is, with **pip install jupyter**, but fortunately, it comes pre-installed with Anaconda. To open a notebook, simply run the command **jupyter notebook** in the Terminal or Command Prompt.

ACCESSING THE CODE FILES

You can find the complete code files of this book at https://packt.live/2DnXRLS. You can also run many activities and exercises directly in your web browser by using the interactive lab environment at https://packt.live/39dH7ml.

We've tried to support interactive versions of all activities and exercises, but we recommend a local installation as well for instances where this support isn't available.

If you have any issues or questions about installation, please email us at **workshops@packt.com**.

1

INTRODUCTION TO NEURAL NETWORKS AND DEEP LEARNING

OVERVIEW

In this chapter, we will cover the basics of neural networks and how to set up a deep learning programming environment. We will also explore the common components and essential operations of a neural network . We will conclude this chapter with an exploration of a trained neural network created using TensorFlow. By the end of this chapter, you will be able to train a neural network.

INTRODUCTION

This chapter is about understanding what neural networks can do rather than the finer workings of deep learning. For this reason, we will not cover the mathematical concepts underlying deep learning algorithms but will describe the essential pieces that make up a deep learning system and the role of neural networks within that system. We will also look at examples where neural networks have been used to solve real-world problems using these algorithms.

At its core, this chapter challenges you to think about your problem as a mathematical representation of ideas. By the end of this chapter, you will be able to think about a problem as a collection of these representations and to recognize how these representations can be learned by deep learning algorithms.

WHAT ARE NEURAL NETWORKS?

A **neural network** is a network of neurons. In our brain, we have a network of billions of neurons that are interconnected with each other. The neuron is one of the basic elements of the nervous system. The primary function of the neuron is to perform actions as a response to an event and transmit messages to other neurons. In this case, the action is simply either activating or deactivating itself. Taking inspiration from the brain's design, **artificial neural networks** were first proposed in the 1940s by MIT professors *Warren McCullough* and *Walter Pitts*.

> **NOTE**
>
> For more information on neural networks, refer to *Explained: Neural networks. MIT News Office, April 14, 2017*, available at http://news.mit.edu/2017/explained-neural-networks-deep-learning-0414.

Inspired by advancements in neuroscience, they proposed to create a computer system that reproduced how the brain works (human or otherwise). At its core was the idea of a computer system that worked as an interconnected network, that is, a system that has many simple components. These components interpret data and influence each other on how to interpret that data. The same core idea remains today.

Deep learning is largely considered the contemporary study of neural networks. Think of it as a contemporary name given to neural networks. The main difference is that the neural networks used in deep learning are typically far greater in size, meaning they have many more nodes and layers than earlier neural networks. Deep learning algorithms and applications typically require resources to achieve success, hence the use of the word *deep* to emphasize their size and the large number of interconnected components.

SUCCESSFUL APPLICATIONS OF NEURAL NETWORKS

Neural networks have been under research in one form or another since their inception in the 1940s. It is only recently that deep learning systems have been used successfully in large-scale industry applications.

Contemporary proponents of neural networks have demonstrated great success in speech recognition, language translation, image classification, and other fields. Its current prominence is backed by a significant increase in available computing power and the emergence of **Graphic Processing Units** (**GPUs**) and **Tensor Processing Units** (**TPUs**), which can perform many more simultaneous mathematical operations than regular CPUs, as well as much greater availability of data. Compared to CPUs, GPUs are designed to execute special tasks (in the "single instruction, multiple threads" model) where the execution can be parallelized.

One such success story is the power consumption of different AlphaGo algorithms. **AlphaGo** is an initiative by DeepMind to develop a series of algorithms to beat the game Go. It is considered a prime example of the power of deep learning. The team at DeepMind was able to do this using reinforcement learning in which AlphaGo becomes its own teacher.

The neural network, which initially knows nothing, plays with itself to understand which moves lead to victory. The algorithm used TPUs for training. TPUs are a type of chipset developed by Google that are specialized for use in deep learning programs. The article *Alpha Zero: Starting from scratch*, https://deepmind.com/blog/alphago-zero-learning-scratch/, depicts the number of GPUs and TPUs used to train different versions of the AlphaGo algorithm.

> **NOTE**
>
> In this book, we will not be using GPUs to fulfill our activities. GPUs are not required to work with neural networks. In several simple examples—like the ones provided in this book—all computations can be performed using a simple laptop's CPU. However, when dealing with very large datasets, GPUs can be of great help given that the long time taken to train a neural network would otherwise be impractical.

Here are a few examples where neural networks have had a significant impact:

Translating text: In 2017, Google announced the release of a new algorithm for its translation service called **Transformer**. The algorithm consisted of a recurrent neural network called **Long Short-term Memory** (**LSTM**) that was trained to use bilingual text. LSTM is a form of neural network that is applied to text data. Google showed that its algorithm had gained notable accuracy when compared to the industry standard, **Bilingual Evaluation Understudy** (**BLEU**), and was also computationally efficient. BLEU is an algorithm for evaluating the performance of machine-translated text. For more information on this, refer to the Google Research Blog, *Transformer: A Novel Neural Network Architecture for Language Understanding,* August 31, 2017, available at https://research.googleblog.com/2017/08/transformer-novel-neural-network.html.

Self-driving vehicles: Uber, NVIDIA, and Waymo are believed to be using deep learning models to control different vehicle functions related to driving. Each company is researching several possibilities, including training the network using humans, simulating vehicles driving in virtual environments, and even creating a small city-like environment in which vehicles can be trained based on expected and unexpected events.

> **NOTE**
>
> To know more about each of these achievements, refer to the following references.
>
> **Uber:** *Uber's new AI team is looking for the shortest route to self-driving cars, Dave Gershgorn, Quartz, December 5, 2016*, available at https://qz.com/853236/ubers-new-ai-team-is-looking-for-the-shortest-route-to-self-driving-cars/.
>
> **NVIDIA**: *End-to-End Deep Learning for Self-Driving Cars, August 17, 2016*, available at https://devblogs.nvidia.com/deep-learning-self-driving-cars/.
>
> **Waymo**: *Inside Waymo's Secret World for Training Self-Driving Cars. The Atlantic, Alexis C. Madrigal, August 23, 2017*, available at https://www.theatlantic.com/technology/archive/2017/08/inside-waymos-secret-testing-and-simulation-facilities/537648/.

Image recognition: Facebook and Google use deep learning models to identify entities in images and automatically tag these entities as persons from a set of contacts. In both cases, the networks are trained with previously tagged images as well as with images from the target friend or contact. Both companies report that the models can suggest a friend or contact with a high level of accuracy in most cases.

While there are many more examples in other industries, the application of deep learning models is still in its infancy. Many successful applications are yet to come, including the ones that you create.

WHY DO NEURAL NETWORKS WORK SO WELL?

Why are neural networks so powerful? Neural networks are powerful because they can be used to predict any given function with reasonable approximation. If we can represent a problem as a mathematical function and we have data that represents that function correctly, a deep learning model can, given enough resources, be able to approximate that function. This is typically called the *Universal Approximation Theorem*. For more information, refer to Michael Nielsen: *Neural Networks and Deep Learning: A visual proof that neural nets can compute any function*, available at http://neuralnetworksanddeeplearning.com/chap4.html.

We will not be exploring mathematical proofs of the universality principle in this book. However, two characteristics of neural networks should give you the right intuition on how to understand that principle: representation learning and function approximation.

> **NOTE**
>
> For more information, refer to *A Brief Survey of Deep Reinforcement Learning, Kai Arulkumaran, Marc Peter Deisenroth, Miles Brundage, and Anil Anthony Bharath, arXiv, September 28, 2017*, available at https://www.arxiv-vanity.com/papers/1708.05866/.

REPRESENTATION LEARNING

The data used to train a neural network contains representations (also known as *features*) that explain the problem you are trying to solve. For instance, if we are interested in recognizing faces from images, the color values of each pixel from a set of images that contain faces will be used as a starting point. The model will then continuously learn higher-level representations by combining pixels together as it goes through its training process. A pictorial depiction is displayed here:

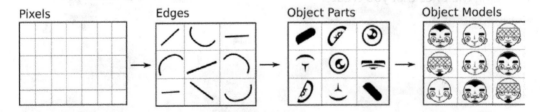

Figure 1.1: A series of higher-level representations based on input data

> **NOTE**
>
> *Figure 1.1* is a derivate image based on an original image from Yann LeCun, Yoshua Bengio, and Geoffrey Hinton in *Deep Learning*, published in *Nature, 521, 436–444 (28 May 2015) doi:10.1038/ nature14539*. You can find the paper at: https://www.nature.com/articles/nature14539.

In formal words, neural networks are computation graphs in which each step computes higher abstraction representations from input data. Each of these steps represents a progression into a different abstraction layer. Data progresses through each of these layers, thereby building higher-level representations. The process finishes with the highest representation possible: the one the model is trying to predict.

FUNCTION APPROXIMATION

When neural networks learn new representations of data, they do so by combining weights and biases with neurons from different layers. They adjust the weights of these connections every time a training cycle occurs using a mathematical technique called **backpropagation**. The weights and biases improve at each round, up to the point that an optimum is achieved. This means that a neural network can measure how wrong it is on every training cycle, adjust the weights and biases of each neuron, and try again. If it determines that a certain modification produces better results than the previous round, it will invest in that modification until an optimal solution is achieved.

So basically, in a single cycle, three things happen. The first one is forward propagation where we calculate the results using weights, biases, and inputs. In the second step, we calculate how far the calculated value is from the expected value using a loss function. The final step is to update the weights and biases moving in the reverse direction of forward propagation, which is called backpropagation.

Since the weights and biases in the earlier layers do not have a direct connection with the later layers, we use a mathematical tool called the chain rule to calculate new weights for the earlier layers. Basically, the change in the earlier layer is equal to the multiplication of the gradients or derivatives of all the layers below it.

In a nutshell, that procedure is the reason why neural networks can approximate functions. However, there are many reasons why a neural network may not be able to predict a function with perfection, chief among them being the following:

- Many functions contain stochastic properties (that is, random properties).

- There may be overfitting to peculiarities from the training data. Overfitting is a situation where the model we are training doesn't generalize well to data it has never seen before. It just learns the training data instead of finding some interesting patterns.

- There may be a lack of training data.

In many practical applications, simple neural networks can approximate a function with reasonable precision. These sorts of applications will be our focus throughout this book.

LIMITATIONS OF DEEP LEARNING

Deep learning techniques are best suited to problems that can be defined with formal mathematical rules (such as data representations). If a problem is hard to define this way, it is likely that deep learning will not provide a useful solution. Moreover, if the data available for a given problem is either biased or only contains partial representations of the underlying functions that generate that problem, deep learning techniques will only be able to reproduce the problem and not learn to solve it.

Remember that deep learning algorithms learn different representations of data to approximate a given function. If data does not represent a function appropriately, it is likely that the function will be incorrectly represented by the neural network. Consider the following analogy: you are trying to predict the national prices of gasoline (that is, fuel) and create a deep learning model. You use your credit card statement with your daily expenses on gasoline as an input data for that model. The model may eventually learn the patterns of your gasoline consumption, but it will likely misrepresent price fluctuations of gasoline caused by other factors only represented weekly in your data such as government policies, market competition, international politics, and so on. The model will ultimately yield incorrect results when used in production.

To avoid this problem, make sure that the data used to train a model represents the problem the model is trying to address as accurately as possible.

INHERENT BIAS AND ETHICAL CONSIDERATIONS

Researchers have suggested that the use of deep learning models without considering the inherent bias in the training data can lead not only to poorly performing solutions but also to ethical complications.

For instance, in late 2016, researchers from the Shanghai Jiao Tong University in China created a neural network that correctly classified criminals using only pictures of their faces. The researchers used 1,856 images of Chinese men, of which half had been convicted. Their model identified inmates with 89.5% accuracy.

NOTE

To know more about this, refer to https://blog.keras.io/the-limitations-of-deep-learning.html and *MIT Technology Review. Neural Network Learns to Identify Criminals by Their Faces, November 22, 2016*, available at https://www.technologyreview.com/2016/11/22/107128/neural-network-learns-to-identify-criminals-by-their-faces/.

The paper resulted in a great furor within the scientific community and popular media. One key issue with the proposed solution is that it fails to properly recognize the bias inherent in the input data. Namely, the data used in this study came from two different sources: one for criminals and one for non-criminals. Some researchers suggest that their algorithm identifies patterns associated with the different data sources used in the study instead of identifying relevant patterns from people's faces. While there are technical considerations that we can make about the reliability of the model, the key criticism is on ethical grounds: researchers ought to clearly recognize the inherent bias in input data used by deep learning algorithms and consider how its application will impact on people's lives. *Timothy Revell. Concerns as face recognition tech used to 'identify' criminals. New Scientist. December 1, 2016. Available at:* https://www.newscientist.com/article/2114900-concerns-as-face-recognition-tech-used-to-identify-criminals/.

There are different types of biases that can occur in a dataset. Consider a case where you are building an automatic surveillance system that can operate both in the daytime and nighttime. So, if your dataset just included images from the daytime, then you would be introducing a sample bias in the model. This could be eliminated by including nighttime data and covering all the different types of cases possible, such as images from a sunny day, a rainy day, and so on. Another example to consider is where, let's suppose, a similar kind of system is installed in a workplace to analyze the workers and their activities. Now, if your model has been fed with thousands of examples with men coding and women cooking, then this data clearly reflects stereotypes. A solution to this problem is the same as earlier: to expose the model to data that is more evenly distributed.

> **NOTE**
>
> To find out more about the topic of ethics in learning algorithms (including deep learning), refer to the work done by the AI Now Institute (https://ainowinstitute.org/), an organization created for the understanding of the social implications of intelligent systems.

COMMON COMPONENTS AND OPERATIONS OF NEURAL NETWORKS

Neural networks have two key components: layers and nodes.

Nodes are responsible for specific operations, and layers are groups of nodes that differentiate different stages of the system. Typically, neural networks are comprised of the following three layers:

- **Input layer**: Where the input data is received and interpreted
- **Hidden layer**: Where computations take place, modifying the data as it passes through
- **Output layer**: Where the output is assembled and evaluated

The following figure displays the working of layers of neural networks:

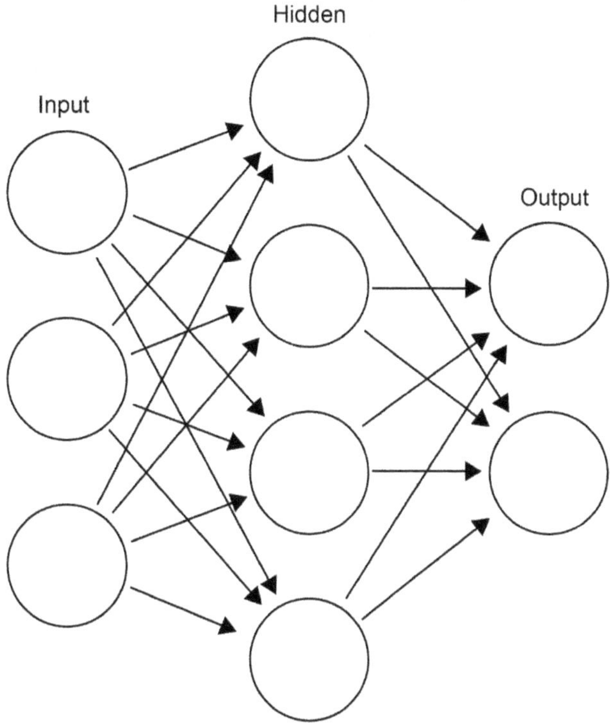

Figure 1.2: An illustration of the most common layers in a neural network

Hidden layers are the most important layers in neural networks. They are referred to as *hidden* because the representations generated in them are not available in the data, but are learned from it instead. It is within these layers where the main computations take place in neural networks.

Nodes are where data is represented in the network. There are two values associated with nodes: biases and weights. Both values affect how data is represented by the nodes and passed on to other nodes. When a network *learns*, it effectively adjusts these values to satisfy an optimization function.

Most of the work in neural networks happens in the hidden layers. Unfortunately, there isn't a clear rule for determining how many layers or nodes a network should have. When implementing a neural network, you will probably spend time experimenting with different combinations of layers and nodes. It is advisable to always start with a single layer and also with a number of nodes that reflect the number of features the input data has (that is, how many *columns* are available in a given dataset).

You can continue to add layers and nodes until a satisfactory performance is achieved—or whenever the network starts overfitting to the training data. Also, note that this depends very much on the dataset – if you were training a model to recognize hand-drawn digits, then a neural network with two hidden layers would be enough, but if your dataset was more complex, say for detecting objects like cars and ambulance in images, then even 10 layers would not be enough and you would need to have a deeper network for the objects to be recognized correctly.

Likewise, if you were using a network with 100 hidden layers for training on handwritten digits, then there would be a strong possibility that you would overfit the model, as that much complexity would not be required by the model in the first place.

Contemporary neural network practice is generally restricted to experimentation with the number of nodes and layers (for example, how deep the network is), and the kinds of operations performed at each layer. There are many successful instances in which neural networks outperformed other algorithms simply by adjusting these parameters.

To start off with, think about data entering a neural network system via the input layer, and then moving through the network from node to node. The path that data takes will depend on how interconnected the nodes are, the weights and the biases of each node, the kind of operations that are performed in each layer, and the state of the data at the end of such operations. Neural networks often require many **runs** (or epochs) in order to keep tuning the weights and biases of nodes, meaning that data flows over the different layers of the graph multiple times.

CONFIGURING A DEEP LEARNING ENVIRONMENT

Before we finish this chapter, we want you to interact with a real neural network. We will start by covering the main software components used throughout this book and make sure that they are properly installed. We will then explore a pre-trained neural network and explore a few of the components and operations discussed in the *What are Neural Networks?* section.

SOFTWARE COMPONENTS FOR DEEP LEARNING

We'll use the following software components for deep learning:

PYTHON 3

We will be using Python 3 in this book. Python is a general-purpose programming language that is very popular with the scientific community—hence its adoption in deep learning. Python 2 is not supported in this book but can be used to train neural networks instead of Python 3. Even if you chose to implement your solutions in Python 2, consider moving to Python 3 as its modern feature set is far more robust than that of its predecessor.

TENSORFLOW

TensorFlow is a library used for performing mathematical operations in the form of graphs. TensorFlow was originally developed by Google, and today, it is an open source project with many contributors. It has been designed with neural networks in mind and is among the most popular choices when creating deep learning algorithms.

TensorFlow is also well known for its production components. It comes with TensorFlow Serving (https://github.com/tensorflow/serving), a high-performance system for serving deep learning models. Also, trained TensorFlow models can be consumed in other high-performance programming languages such as Java, Go, and C. This means that you can deploy these models on anything from a micro-computer (that is, a Raspberry Pi) to an Android device. As of November 2019, TensorFlow version 2.0 is the latest version.

KERAS

In order to interact efficiently with TensorFlow, we will be using Keras (https://keras.io/), a Python package with a high-level API for developing neural networks. While TensorFlow focuses on components that interact with each other in a computational graph, Keras focuses specifically on neural networks. Keras uses TensorFlow as its backend engine and makes developing such applications much easier.

As of November 2019, Keras is the built-in and default API of TensorFlow. It is available under the **tf.keras** namespace.

TENSORBOARD

TensorBoard is a data visualization suite for exploring TensorFlow models and is natively integrated with TensorFlow. TensorBoard works by consuming the checkpoint and summary files created by TensorFlow as it trains a neural network. Those can be explored either in near real time (with a 30-second delay) or after the network has finished training. TensorBoard makes the process of experimenting with and exploring a neural network much easier—plus, it's quite exciting to follow the training of your network.

JUPYTER NOTEBOOK, PANDAS, AND NUMPY

When working to create deep learning models with Python, it is common to start working interactively; slowly developing a model that eventually turns into more structured software. Three Python packages are used frequently during this process: Jupyter Notebooks, Pandas, and NumPy:

- Jupyter Notebook create interactive Python sessions that use a web browser as their interface.

- Pandas is a package for data manipulation and analysis.

- NumPy is frequently used for shaping data and performing numerical computations.

These packages are used occasionally throughout this book. They typically do not form part of a production system but are often used when exploring data and starting to build a model. We'll focus on the other tools in much more detail.

> **NOTE**
>
> The books *Learning pandas* by Michael Heydt (June 2017, Packt Publishing), available at https://www.packtpub.com/big-data-and-business-intelligence/learning-pandas-second-edition, and *Learning Jupyter* by Dan Toomey (November 2016, Packt Publishing), available at https://www.packtpub.com/big-data-and-business-intelligence/learning-jupyter-5-second-edition, both offer comprehensive guides on how to use these technologies. These books are good references for continuing to learn more.

The following table details the software requirements required for successfully creating the deep learning models explained in this book:

Component	Description	Minimum Version
Python	A general-purpose programming language. A popular language used in the development of deep learning applications.	3.7
TensorFlow	An open source graph computation Python package typically used for developing deep learning systems.	2.0
Keras	A Python package that provides a high-level interface to TensorFlow.	2.2.4
TensorBoard	Browser-based software for visualizing neural network statistics.	2.0.1
Jupyter Notebook	Browser-based software for working interactively with Python sessions.	1.0.0
pandas	A Python package for analyzing and manipulating data.	0.21.0
NumPy	A Python package for high-performance numerical computations.	1.13.3

Figure 1.3: The software components necessary for creating a deep learning environment

Anaconda is a free distribution of many useful Python packages for Windows, Mac or other platform. We recommend that you follow the instructions at https://docs.anaconda.com/anaconda/install/. The standard Anaconda installation will install most of these components and the first exercise will work through how to install the others.

EXERCISE 1.01: VERIFYING THE SOFTWARE COMPONENTS

Before we explore a trained neural network, let's verify whether all the software components that we need are available. We have included a script that verifies whether these components work. Let's take a moment to run the script and deal with any eventual problems we may find. We will now be testing whether the software components required for this book are available in your working environment. First, we suggest the creation of a Python virtual environment using Python's native module **venv**. Virtual environments are used for managing project dependencies. We suggest each project you create has its own virtual environment.

1. A python virtual environment can be created by using the following command:

```
$ python -m venv venv
$ source venv/bin/activate
```

The latter command will append the string **venv** at the beginning of the command line.

Make sure you always activate your Python virtual environment when working on a project. To deactivate your virtual environment, run **$ deactivate**.

2. After activating your virtual environment, make sure that the right components are installed by executing **pip** over the **requirements.txt** file (https://packt.live/300skHu).

```
$ pip install -r requirements.txt
```

The output is as follows:

Figure 1.4: A screenshot of a Terminal running pip to install dependencies from requirements.txt

3. This will install the libraries used in this book in that virtual environment. It will do nothing if they are already available. If the library is getting installed, a progress bar will be shown, else it will notify that '`requirement is already specified`'. To check the available libraries installed, please use the following command:

```
$ pip list
```

The output will be as follows:

```
(trash) (base) MacBook-Pro-3:Exercise1.01 alexisrutherford$ pip list
Package                Version
--------------------   ---------
absl-py                0.8.1
appnope                0.1.0
astor                  0.8.0
attrs                  19.3.0
backcall               0.1.0
beautifulsoup4         4.8.1
bleach                 3.1.0
cachetools             3.1.1
certifi                2019.9.11
chardet                3.0.4
Click                  7.0
cycler                 0.10.0
decorator              4.4.1
defusedxml             0.6.0
entrypoints            0.3
Flask                  1.1.1
Flask-API              2.0
Flask-Caching          1.8.0
Flask-Cors             3.0.8
```

Figure 1.5: A screenshot of a Terminal running pip to list the available libraries

NOTE

These libraries are essential for working with all the code activities in this book.

4. As a final step in this exercise, execute the script **test_stack.py**. This can be found at: https://packt.live/2B0JNau It verifies that all the required packages for this book are installed and available in your system.

5. Run the following script to check if the dependencies of Python 3, TensorFlow, and Keras are available. Use the following command:

```
$ python3 Chapter01/Exercise1.01/test_stack.py
```

The script returns helpful messages stating what is installed and what needs to be installed:

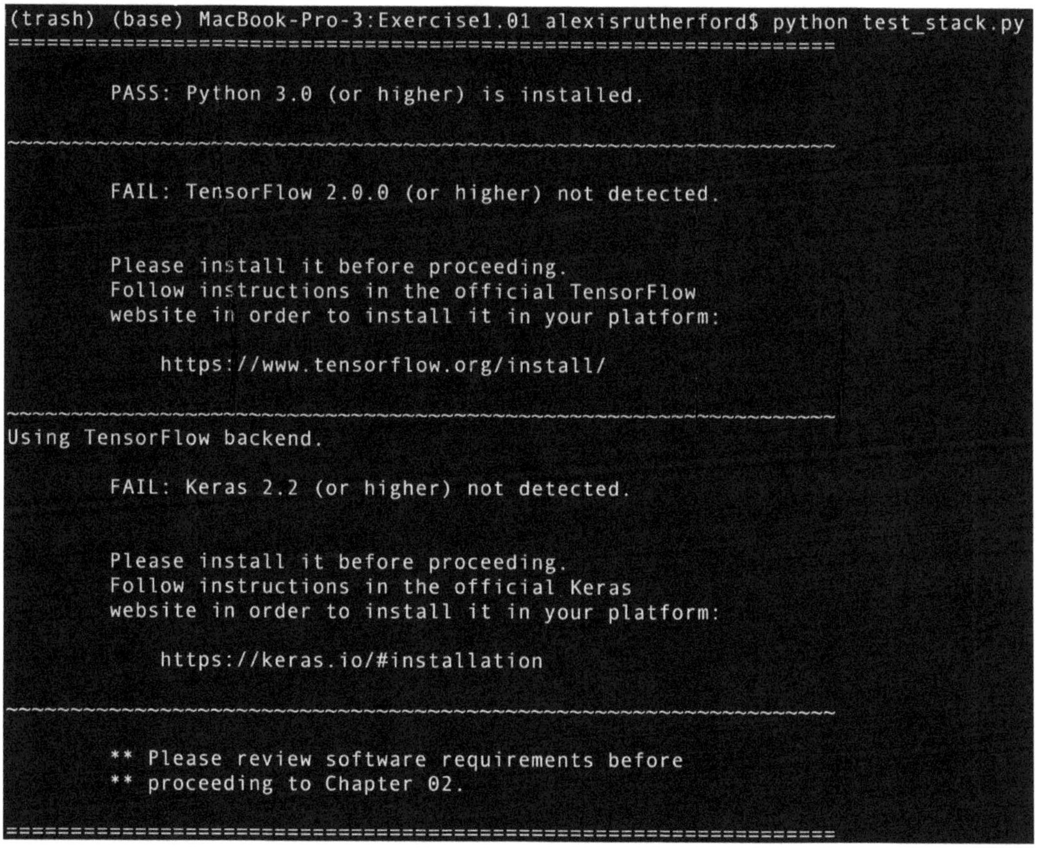

```
(trash) (base) MacBook-Pro-3:Exercise1.01 alexisrutherford$ python test_stack.py
================================================================
        PASS: Python 3.0 (or higher) is installed.


        FAIL: TensorFlow 2.0.0 (or higher) not detected.


        Please install it before proceeding.
        Follow instructions in the official TensorFlow
        website in order to install it in your platform:

            https://www.tensorflow.org/install/

Using TensorFlow backend.

        FAIL: Keras 2.2 (or higher) not detected.


        Please install it before proceeding.
        Follow instructions in the official Keras
        website in order to install it in your platform:

            https://keras.io/#installation


        ** Please review software requirements before
        ** proceeding to Chapter 02.
================================================================
```

Figure 1.6: A screenshot of a Terminal displaying that not all the requirements are installed

For example, in the preceding screenshot, it shows that TensorFlow 2.0 is not detected but Keras 2.2 or higher is detected. Hence you are shown the error message **Please review software requirements before proceeding to Lesson 2**. If all the requirements are fulfilled, then it will show Python, TensorFlow, and Keras as installed, as shown in the following screenshot:

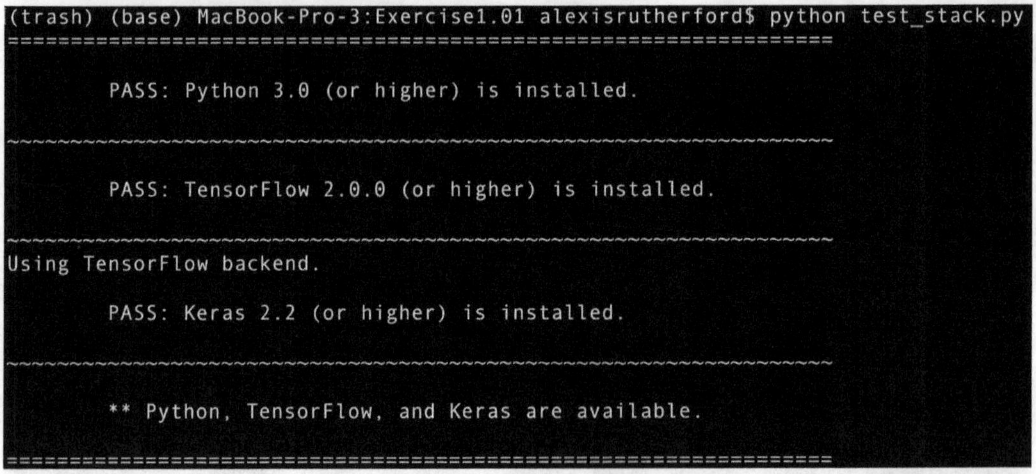

```
(trash) (base) MacBook-Pro-3:Exercise1.01 alexisrutherford$ python test_stack.py
================================================================

        PASS: Python 3.0 (or higher) is installed.


~~~~~~~~~~~~~~~~~~~~~~~~~~~~~~~~~~~~~~~~~~~~~~~~~~~~~~~~~~~~~~~~~~~~

        PASS: TensorFlow 2.0.0 (or higher) is installed.

~~~~~~~~~~~~~~~~~~~~~~~~~~~~~~~~~~~~~~~~~~~~~~~~~~~~~~~~~~~~~~~~~~~~
Using TensorFlow backend.
        PASS: Keras 2.2 (or higher) is installed.

~~~~~~~~~~~~~~~~~~~~~~~~~~~~~~~~~~~~~~~~~~~~~~~~~~~~~~~~~~~~~~~~~~~~

        ** Python, TensorFlow, and Keras are available.

================================================================
```

Figure 1.7: A screenshot of the Terminal displaying that all elements are installed

6. Run the following script command in your Terminal to find more information on how to configure TensorBoard:

```
$ tensorboard -help
```

The output is as follows:

```
(trash) (base) MacBook-Pro-3:Exercise1.01 alexisrutherford$ tensorboard --help
usage: tensorboard [-h] [--helpfull] [--logdir PATH] [--logdir_spec PATH_SPEC]
                   [--host ADDR] [--bind_all] [--port PORT]
                   [--purge_orphaned_data BOOL] [--db URI] [--db_import]
                   [--inspect] [--version_tb] [--tag TAG] [--event_file PATH]
                   [--path_prefix PATH] [--window_title TEXT]
                   [--max_reload_threads COUNT] [--reload_interval SECONDS]
                   [--reload_task TYPE] [--reload_multifile BOOL]
                   [--reload_multifile_inactive_secs SECONDS]
                   [--generic_data TYPE]
                   [--samples_per_plugin SAMPLES_PER_PLUGIN]
                   [--debugger_data_server_grpc_port PORT]
                   [--debugger_port PORT] [--master_tpu_unsecure_channel ADDR]
                   {serve,dev} ...

TensorBoard is a suite of web applications for inspecting and understanding
your TensorFlow runs and graphs. https://github.com/tensorflow/tensorboard

positional arguments:
  {serve,dev}           TensorBoard subcommand (defaults to 'serve')
    serve               start local TensorBoard server (default subcommand)
    dev                 upload data to TensorBoard.dev

optional arguments:
  -h, --help            show this help message and exit
  --helpfull            show full help message and exit
  --logdir PATH         Directory where TensorBoard will look to find
                        TensorFlow event files that it can display.
                        TensorBoard will recursively walk the directory
                        structure rooted at logdir, looking for .*tfevents.*
                        files. A leading tilde will be expanded with the
                        semantics of Python's os.expanduser function.
```

Figure 1.8: An output of the --help command

You should see the relevant help messages that explain what each command does, as in *Figure 1.8*.

As you can see in the figure above, the script returns messages informing you that all dependencies are installed correctly.

> **NOTE**
>
> To access the source code for this specific section, please refer to https://packt.live/2B0JNau.
>
> This section does not currently have an online interactive example, and will need to be run locally.

Once we have verified that Python 3, TensorFlow, Keras, TensorBoard, and the packages outlined in **requirements.txt** have been installed, we can continue to a demo on how to train a neural network and then go on to explore a trained network using these same tools.

EXPLORING A TRAINED NEURAL NETWORK

In this section, we'll explore a trained neural network. We'll do this to understand how a neural network solves a real-world problem (predicting handwritten digits) and to get familiar with the TensorFlow API. When exploring this neural network, we will recognize many components introduced in previous sections, such as nodes and layers, but we will also see many that we don't recognize (such as activation functions); we will explore those in further sections. We will then walk through an exercise on how that neural network was trained and then train that same network ourselves.

The network that we will be exploring has been trained to recognize numerical digits (integers) using images of handwritten digits. It uses the MNIST dataset (http://yann.lecun.com/exdb/mnist/), a classic dataset frequently used for exploring pattern recognition tasks.

THE MNIST DATASET

The **Modified National Institute of Standards and Technology** (**MNIST**) dataset contains a training set of 60,000 images and a test set of 10,000 images. Each image contains a single handwritten number. This dataset, which is derived from one created by the US Government, was originally used to test different approaches to the problem of recognizing handwritten text by computer systems. Being able to do that was important for the purpose of increasing the performance of postal services, taxation systems, and government services. The MNIST dataset is considered too naïve for contemporary methods. Different and more recent datasets are used in contemporary research (for example, **Canadian Institute for Advanced Research** (**CIFAR**). However, the MNIST dataset is still very useful for understanding how neural networks work because known models can achieve a high level of accuracy with great efficiency.

> **NOTE**
>
> The CIFAR dataset is a machine learning dataset that contains images organized in different classes. Different than the MNIST dataset, the CIFAR dataset contains classes from many different areas including animals, activities, and objects. The CIFAR dataset is available at https://www. cs.toronto.edu/~kriz/cifar.html.

However, the MNIST dataset is still very useful for understanding how neural networks work because known models can achieve a high level of accuracy with great efficiency. In the following figure, each image is a separate 20x20-pixel image containing a single handwritten digit. You can find the original dataset at http://yann. lecun.com/exdb/mnist/.

Figure 1.9: An excerpt from the training set of the MNIST dataset

TRAINING A NEURAL NETWORK WITH TENSORFLOW

Now, let's train a neural network to recognize new digits using the MNIST dataset. We will be implementing a special-purpose neural network called a **Convolutional Neural Network(CNN)** to solve this problem (we will discuss those in more detail in later sections). Our complete network contains three hidden layers: two fully connected layers and a convolutional layer. The model is defined by the following TensorFlow snippet of Python code:

> **NOTE**
>
> The code snippet shown here uses a backslash (\ ) to split the logic across multiple lines. When the code is executed, Python will ignore the backslash, and treat the code on the next line as a direct continuation of the current line.

```
model = Sequential()
model.add(Convolution2D(filters = 10, kernel_size = 3, \
                        input_shape=(28,28,1)))
model.add(Flatten())
model.add(Dense(128, activation = 'relu'))
model.add(Dropout(0.2))
model.add(Dense(10, activation = 'softmax'))
```

> **NOTE**
>
> Use the **mnist.py** file for your reference at https://packt.live/2Cuhj9w. Follow along by opening the script in your code editor.

We execute the preceding snippet of code only once during the training of our network.

We will go into a lot more detail about each one of those components using Keras in *Chapter 2, Real-World Deep Learning: Predicting the Price of Bitcoin*. For now, we'll focus on understanding that the network is altering the values of the **Weights** and **Biases** in each layer on every run. These lines of Python are the culmination of dozens of years of neural network research.

Now let's train that network to evaluate how it performs in the MNIST dataset.

EXERCISE 1.02: TRAINING A NEURAL NETWORK USING THE MNIST DATASET

In this exercise, we will train a neural network for detecting handwritten digits from the MNIST dataset. Execute the following steps to set up this exercise:

1. Open two Terminal instances.

2. Navigate to https://packt.live/2BWNAWK. Ensure that your Python 3 virtual environment is active and that the requirements outlined in **requirements. txt** are installed.

3. In one of them, start a TensorBoard server with the following command:

```
$ tensorboard --logdir logs/fit
```

The output is as follows:

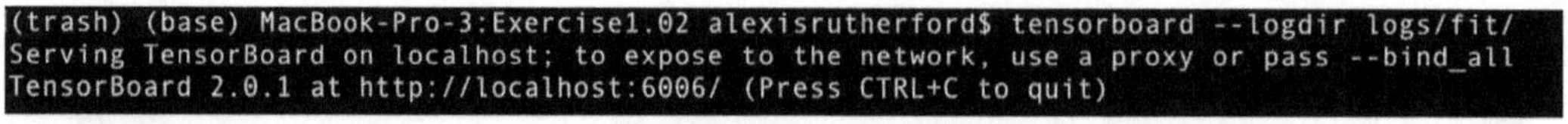

```
(trash) (base) MacBook-Pro-3:Exercise1.02 alexisrutherford$ tensorboard --logdir logs/fit/
Serving TensorBoard on localhost; to expose to the network, use a proxy or pass --bind_all
TensorBoard 2.0.1 at http://localhost:6006/ (Press CTRL+C to quit)
```

Figure 1.10: The TensorBoard server

In the other, run the **mnist.py** script from within that directory with the following command:

```
$ python mnist.py
```

When you start running the script, you will see the progress bar as follows:

```
2020-03-21 13:51:23.805700: I tensorflow/core/platform/cpu_feature_guard.cc:142] Your CPU supports instructions that this Tenso
rFlow binary was not compiled to use: AVX2 FMA
2020-03-21 13:51:23.828279: I tensorflow/compiler/xla/service/service.cc:168] XLA service 0x7fce3a33acb0 executing computations
on platform Host. Devices:
2020-03-21 13:51:23.828298: I tensorflow/compiler/xla/service/service.cc:175]   StreamExecutor device (0): Host, Default Versio
n
Train on 60000 samples, validate on 10000 samples
Epoch 1/5
2020-03-21 13:51:24.422694: I tensorflow/core/profiler/lib/profiler_session.cc:184] Profiler session started.
60000/60000 [==============================] - 16s 275us/sample - loss: 0.2429 - accuracy: 0.9272 - val_loss: 0.1073 - val_accu
racy: 0.9682
Epoch 2/5
12544/60000 [=====>........................] - ETA: 12s - loss: 0.1405 - accuracy: 0.9584
```

Figure 1.11: The result of the mnist.py script

4. Open your browser and navigate to the TensorBoard URL provided when you start the server in *step 3*, it might be **http://localhost:6006/** or similar. In the Terminal where you ran the **mnist.py** script, you will see a progress bar with the epochs of the model. When you open the browser page, you will see a couple of graphs, **epoch_accuracy** and **epoch_loss** graphs. Ideally, the accuracy should improve with each iteration and the loss should decrease with each iteration. You can confirm this visually with the graphs.

5. Click the **epoch_accuracy** graph, enlarge it, and let the page refresh (or click on the **refresh** icon). You will see the model gaining accuracy as the epochs go by:

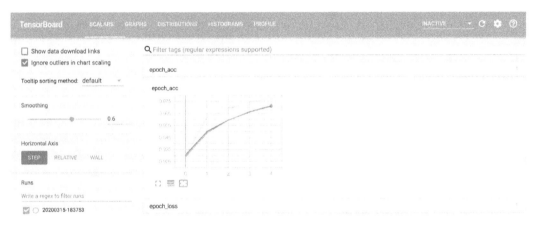

Figure 1.12: A visualization of the accuracy and loss graphs using TensorBoard

We can see that after about 5 epochs (or steps), the network surpassed 97% accuracy. That is, the network is getting 97% of the digits in the test set correct by this point.

> **NOTE**
>
> To access the source code for this specific section, please refer to https://packt.live/2Cuhj9w.
>
> This section does not currently have an online interactive example, and will need to be run locally.

Now, let's also test how well those networks perform with unseen data.

TESTING NETWORK PERFORMANCE WITH UNSEEN DATA

Visit the website http://mnist-demo.herokuapp.com/ in your browser and draw a number between 0 and 9 in the designated white box:

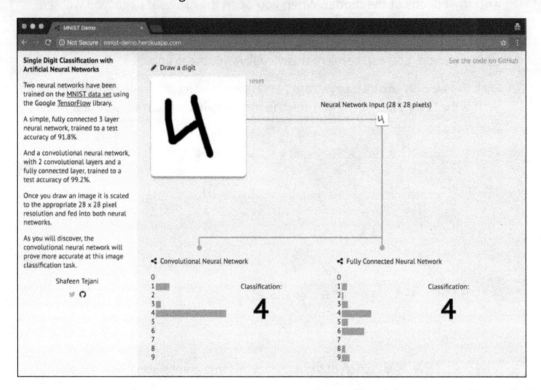

Figure 1.13: A web application for manually drawing digits and testing the accuracy of two trained networks

> **NOTE**
>
> This web application we are using was created by *Shafeen Tejani* to explore whether a trained network can correctly predict handwritten digits that we create.
>
> Source: https://github.com/ShafeenTejani/mnist-demo.

In the application, you can see the results of two neural networks – a **Convolutional Neural Network** (**CNN**) and a **Fully Connected Neural Network**. The one that we have trained is the CNN. Does it classify all your handwritten digits correctly? Try drawing numbers at the edge of the designated area. For instance, try drawing the number **1** close to the right edge of the drawing area, as shown in the following figure:

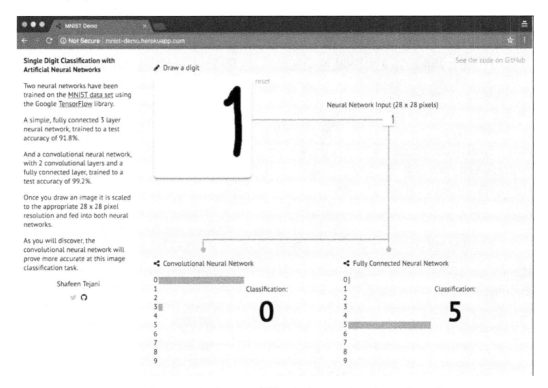

Figure 1.14: Both networks have a difficult time estimating values drawn on the edges of the area

In this example, we see the number 1 drawn to the right side of the drawing area. The probability of this number being a 1 is 0 in both networks.

The MNIST dataset does not contain numbers on the edges of images. Hence, neither network assigns relevant values to the pixels located in that region. Both networks are much better at classifying numbers correctly if we draw them closer to the center of the designated area. This is due to the fact that in the training set, we only had images with numbers drawn in the center of the image. This shows that neural networks can only be as powerful as the data that is used to train them. If the data used for training is very different than what we are trying to predict, the network will most likely produce disappointing results.

ACTIVITY 1.01: TRAINING A NEURAL NETWORK WITH DIFFERENT HYPERPARAMETERS

In this section, we will explore the neural network that we trained during our work on *Exercise 1.02, Training a Neural Network Using the MNIST Dataset*, where we trained our own CNN on the MNIST dataset. We have provided that same trained network as binary files in the directory of this book. In this activity, we will just cover the things that you can do using TensorBoard and we will train several other networks by just changing some hyperparameters.

Here are the steps you need to follow:

1. Open TensorBoard by writing the appropriate command.

2. Open the TensorBoard accuracy graph and play with the values of smoothening sliders in scalar areas.

3. Train another model by changing the hyperparameters.

4. Try decreasing the learning rate and increasing the number of epochs.

5. Now try to understand what effect this hyperparameter tuning has on the graphs generated on TensorBoard.

6. Try increasing the learning rate and decreasing the number of epochs and repeat *step 5*.

> NOTE:
>
> The solution for this activity can be found on page 130.

SUMMARY

In this chapter, we explored a TensorFlow-trained neural network using TensorBoard and trained our own modified version of that network with different epochs and learning rates. This gave you hands-on experience of how to train a highly performant neural network and allowed you to explore some of its limitations.

Do you think we can achieve similar accuracy with real Bitcoin data? We will attempt to predict future Bitcoin prices using a common neural network algorithm in *Chapter 2, Real-World Deep Learning: Predicting the Price of Bitcoin*. In *Chapter 3, Real-World Deep Learning: Evaluating the Bitcoin Model*, we will evaluate and improve that model, and finally, in *Chapter 4, Productization*, we will create a program that serves the prediction of that system via an HTTP API.

2

REAL-WORLD DEEP LEARNING: PREDICTING THE PRICE OF BITCOIN

OVERVIEW

This chapter will help you to prepare data for a deep learning model, choose the right model architecture, use Keras—the default API of TensorFlow 2.0, and make predictions with the trained model. By the end of this chapter, you will have prepared a model to make predictions which we will explore in the upcoming chapters.

INTRODUCTION

Building on fundamental concepts from *Chapter 1, Introduction to Neural Networks and Deep Learning*, let's now move on to a real-world scenario and identify whether we can build a deep learning model that predicts Bitcoin prices.

We will learn the principles of preparing data for a deep learning model, and how to choose the right model architecture. We will use Keras—the default API of TensorFlow 2.0 and make predictions with the trained model. We will conclude this chapter by putting all these components together and building a bare bones, yet complete, first version of a deep learning application.

Deep learning is a field that is undergoing intense research activity. Among other things, researchers are devoted to inventing new neural network architectures that can either tackle new problems or increase the performance of previously implemented architectures.

In this chapter, we will study both old and new architectures. Older architectures have been used to solve a large array of problems and are generally considered the right choice when starting a new project. Newer architectures have shown great success in specific problems but are harder to generalize. The latter are interesting as references of what to explore next but are hardly a good choice when starting a project.

The following topic discusses the details of these architectures and how to determine the best one for a particular problem statement.

CHOOSING THE RIGHT MODEL ARCHITECTURE

Considering the available architecture possibilities, there are two popular architectures that are often used as starting points for several applications: **Convolutional Neural Networks (CNNs)** and **Recurrent Neural Networks (RNNs)**. These are foundational networks and should be considered starting points for most projects.

We also include descriptions of another three networks, due to their relevance in the field: **Long Short-Term Memory (LSTM)** networks (an RNN variant); **Generative Adversarial Networks (GANs)**; and **Deep Reinforcement Learning (DRL)**. These latter architectures have shown great success in solving contemporary problems, however, they are slightly difficult to use. The next section will cover the use of different types of architecture in different problems.

CONVOLUTIONAL NEURAL NETWORKS (CNNS)

CNNs have gained notoriety for working with problems that have a grid-like structure. They were originally created to classify images, but have been used in several other areas, ranging from speech recognition to self-driving vehicles.

A CNN's essential insight is to use closely related data as an element of the training process, instead of only individual data inputs. This idea is particularly effective in the context of images, where a pixel located at the center is related to the ones located to its right and left. The name **convolution** is given to the mathematical representation of the following process:

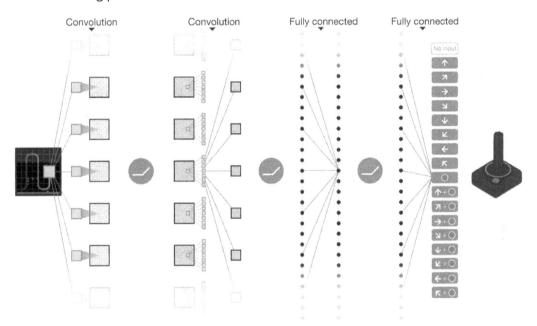

Figure 2.1: Illustration of the convolution process

NOTE

Image source: Volodymyr Mnih, *et al.*

You can find this image at: https://packt.live/3fivWLB

For more information about deep reinforcement learning, refer to *Human-level Control through Deep Reinforcement Learning. February 2015, Nature*, available at https://storage.googleapis.com/deepmind-media/dqn/DQNNaturePaper.pdf.

RECURRENT NEURAL NETWORKS (RNNS)

A CNN works with a set of inputs that keeps altering the weights and biases of the network's respective layers and nodes. A known limitation of this approach is that its architecture ignores the sequence of these inputs when determining the changes to the network's weights and biases.

RNNs were created precisely to address that problem. They are designed to work with sequential data. This means that at every epoch, layers can be influenced by the output of previous layers. The memory of previous observations in each sequence plays an important role in the evaluation of posterior observations.

RNNs have had successful applications in speech recognition due to the sequential nature of that problem. Also, they are used for translation problems. Google Translate's current algorithm—Transformer, uses an RNN to translate text from one language to another. In late 2018, Google introduced another algorithm based on the Transformer algorithm called **Bidirectional Encoder Representations from Transformers (BERT)**, which is currently state of the art in **Natural Language Processing (NLP)**.

> **NOTE**
>
> For more information RNN applications, refer to the following:
>
> *Transformer: A Novel Neural Network Architecture for Language Understanding, Jakob Uszkoreit, Google Research Blog, August 2017,* available at https://ai.googleblog.com/2017/08/transformer-novel-neural-network.html.
>
> *BERT: Open Sourcing BERT: State-of-the-Art Pre-Training for Natural Language Processing*, available at https://ai.googleblog.com/2018/11/open-sourcing-bert-state-of-art-pre.html.

The following diagram illustrates how words in English are linked to words in French, based on where they appear in a sentence. RNNs are very popular in language translation problems:

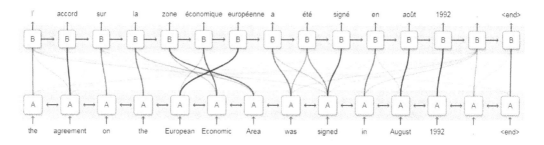

Figure 2.2: Illustration from distill.pub linking words in English and French

> **NOTE**
>
> Image source: https://distill.pub/2016/augmented-rnns/

LONG SHORT-TERM MEMORY (LSTM) NETWORKS

LSTM networks are RNN variants created to address the vanishing gradient problem. This problem is caused by memory components that are too distant from the current step that receive lower weights due to their distance. LSTMs are a variant of RNNs that contain a memory component called a **forget gate**. This component can be used to evaluate how both recent and old elements affect the weights and biases, depending on where the observation is placed in a sequence.

> **NOTE**
>
> The LSTM architecture was first introduced by Sepp Hochreiter and Jürgen Schmidhuber in 1997. Current implementations have had several modifications. For a detailed mathematical explanation of how each component of an LSTM works, refer to the article *Understanding LSTM Networks*, by Christopher Olah, August 2015, available at https://colah.github.io/posts/2015-08-Understanding-LSTMs/.

GENERATIVE ADVERSARIAL NETWORKS

Generative Adversarial Networks (**GANs**) were invented in 2014 by Ian Goodfellow and his colleagues at the University of Montreal. GANs work based on the approach that, instead of having one neural network that optimizes weights and biases with the objective to minimize its errors, there should be two neural networks that compete against each other for that purpose.

> **NOTE**
>
> For more information on GANs, refer to *Generative Adversarial Networks, Ian Goodfellow, et al., arXiv. June 10, 2014*, available at https://arxiv.org/pdf/1406.2661.pdf.

GANs generate new data (*fake* data) and a network that evaluates the likelihood of the data generated by the first network being *real* or *fake*. They compete because both learn: one learns how to better generate *fake* data, and the other learns how to distinguish whether the data presented is real. They iterate on every epoch until convergence. That is the point when the network that evaluates generated data cannot distinguish between *fake* and *real* data any further.

GANs have been successfully used in fields where data has a clear topological structure. Originally, GANs were used to create synthetic images of objects, people's faces, and animals that were similar to real images of those things. You will see in the following image, used in the **StarGAN** project, that the expressions on the face change:

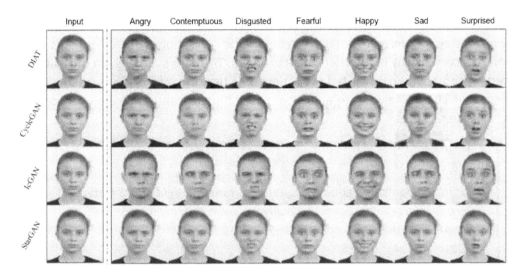

Figure 2.3: Changes in people's faces based on emotion, using GAN algorithms

This domain of image creation is where GANs are used the most frequently, but applications in other domains occasionally appear in research papers.

NOTE

Image source: StarGAN project, available at https://github.com/yunjey/StarGAN.

DEEP REINFORCEMENT LEARNING (DRL)

The original DRL architecture was championed by DeepMind, a Google-owned artificial intelligence research organization based in the UK. The key idea of DRL networks is that they are unsupervised in nature and that they learn from trial and error, only optimizing for a reward function.

That is, unlike other networks (which use a supervised approach to optimize incorrect predictions as compared to what are known to be correct ones), DRL networks do not know of a correct way of approaching a problem. They are simply given the rules of a system and are then rewarded every time they perform a function correctly. This process, which takes a very large number of iterations, eventually trains networks to excel in several tasks.

> **NOTE**
>
> For more information about DRL, refer to *Human-Level Control through Deep Reinforcement Learning, Volodymyr Mnih et al., February 2015, Nature,* available at: https://storage.googleapis.com/deepmind-media/dqn/DQNNaturePaper.pdf.

DRL models gained popularity after DeepMind created AlphaGo—a system that plays the game Go better than professional players. DeepMind also created DRL networks that learn how to play video games at a superhuman level, entirely on their own.

> **NOTE**
>
> For more information about DQN, look up the DQN that was created by DeepMind to beat Atari games. The algorithm uses a DRL solution to continuously increase its reward.
>
> Image source: https://keon.io/deep-q-learning/.

Here's a summary of neural network architectures and their applications:

Architecture	Data Structure	Successful Applications
CNNs	Grid-like topological structure (that is, images)	Image recognition and classification
RNNs and LSTM networks	Sequential data (that is, time series data)	Speech recognition, text generation, and translation
GANs	Grid-like topological structure (that is, images)	Image generation
DRL	System with clear rules and a clearly defined reward function	Playing video games and self-driving vehicles

Figure 2.4: Different neural network architectures, data structures, and their successful applications

DATA NORMALIZATION

Before building a deep learning model, data normalization is an important step. Data normalization is a common practice in machine learning systems. For neural networks, researchers have proposed that normalization is an essential technique for training RNNs (and LSTMs), mainly because it decreases the network's training time and increases the network's overall performance.

> **NOTE**
>
> For more information, refer to *Batch Normalization: Accelerating Deep Network Training by Reducing Internal Covariate Shift*, *Sergey Ioffe et al.*, *arXiv*, March 2015, available at https://arxiv.org/abs/1502.03167.

Which normalization technique works best depends on the data and the problem at hand. A few commonly used techniques are listed here:

Z-SCORE

When data is normally distributed (that is, Gaussian), you can compute the distance between each observation as a standard deviation from its mean. This normalization is useful when identifying how distant the data points are from more likely occurrences in the distribution. The Z-score is defined by the following formula:

$$z_i = \frac{x_i - \mu}{\sigma}$$

Figure 2.5: Formula for Z-Score

Here, x_i is the ith observation, μ is the mean, and σ is the standard deviation of the series.

> **NOTE**
>
> For more information, refer to the standard score article (*Z-Score: Definition, Formula, and Calculation*), available at https://www.statisticshowto.datasciencecentral.com/probability-and-statistics/z-score/.

POINT-RELATIVE NORMALIZATION

This normalization computes the difference in a given observation in relation to the first observation of the series. This kind of normalization is useful for identifying trends in relation to a starting point. The point-relative normalization is defined by:

$$n_i = \left(\frac{o_i}{o_0}\right) - 1$$

Figure 2.6: Formula for point-relative normalization

Here, o_i is the ith observation, and o_o is the first observation of the series.

> **NOTE**
>
> For more information on making predictions, watch *How to Predict Stock Prices Easily – Intro to Deep Learning #7, Siraj Raval*, available on YouTube at https://www.youtube.com/watch?v=ftMq5ps503w.

MAXIMUM AND MINIMUM NORMALIZATION

This type of normalization computes the distance between a given observation and the maximum and minimum values of the series. This is useful when working with series in which the maximum and minimum values are not outliers and are important for future predictions. This normalization technique can be applied with the following formula:

$$n_i = \frac{o_i - \min(O)}{\max(O) - \min(O)}$$

Figure 2.7: Formula for calculating normalization

Here, O_i is the ith observation, O represents a vector with all O values, and the functions *min (O)* and *max (O)* represent the minimum and maximum values of the series, respectively.

During *Exercise 2.01, Exploring the Bitcoin Dataset and Preparing Data for a Model*, we will prepare available Bitcoin data to be used in our LSTM model. That includes selecting variables of interest, selecting a relevant period, and applying the preceding point-relative normalization technique.

STRUCTURING YOUR PROBLEM

Compared to researchers, practitioners spend much less time determining which architecture to choose when starting a new deep learning project. Acquiring data that represents a given problem correctly is the most important factor to consider when developing these systems, followed by an understanding of the dataset's inherent biases and limitations. When starting to develop a deep learning system, consider the following questions for reflection:

- Do I have the right data? This is the hardest challenge when training a deep learning model. First, define your problem with mathematical rules. Use precise definitions and organize the problem into either categories (classification problems) or a continuous scale (regression problems). Now, how can you collect data pertaining to those metrics?

- Do I have enough data? Typically, deep learning algorithms have shown to perform much better on large datasets than on smaller ones. Knowing how much data is necessary to train a high-performance algorithm depends on the kind of problem you are trying to address, but aim to collect as much data as you can.

- Can I use a pre-trained model? If you are working on a problem that is a subset of a more general application, but within the same domain. Consider using a pre-trained model. Pre-trained models can give you a head start on tackling the specific patterns of your problem, instead of the more general characteristics of the domain at large. A good place to start is the official TensorFlow repository (https://github.com/tensorflow/models).

When you structure your problem with such questions, you will have a sequential approach to any new deep learning project. The following is a representative flow chart of these questions and tasks:

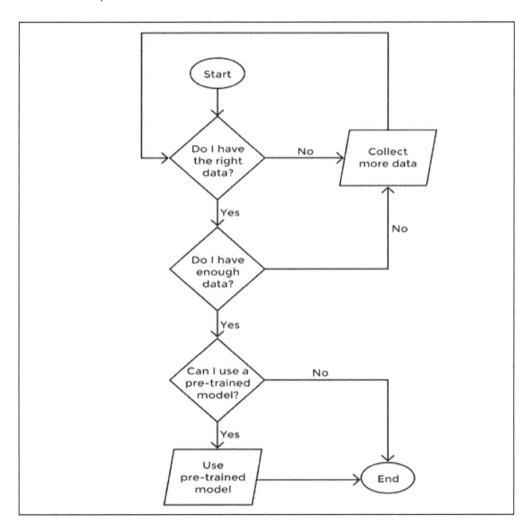

Figure 2.8: Decision tree of key reflection questions to be asked at the beginning of a deep learning project

In certain circumstances, the data may simply not be available. Depending on the case, it may be possible to use a series of techniques to effectively create more data from your input data. This process is known as **data augmentation** and can be applied successfully when working with image recognition problems.

> **NOTE**
>
> A good reference is the article *Classifying plankton with deep neural networks*, available at https://benanne.github.io/2015/03/17/plankton.html. The authors show a series of techniques for augmenting a small set of image data in order to increase the number of training samples the model has.

Once the problem is well-structured, you will be able to start preparing the model.

JUPYTER NOTEBOOK

We will be using Jupyter Notebook to code in this section. Jupyter Notebooks provide Python sessions via a web browser that allows you to work with data interactively. They are a popular tool for exploring datasets. They will be used in exercises throughout this book.

EXERCISE 2.01: EXPLORING THE BITCOIN DATASET AND PREPARING DATA FOR A MODEL

In this exercise, we will prepare the data and then pass it to the model. The prepared data will then be useful in making predictions as we move ahead in the chapter. Before preparing the data, we will do some visual analysis on it, such as looking at when the value of Bitcoin was at its highest and when the decline started.

> **NOTE**
>
> We will be using a public dataset originally retrieved from the Yahoo Finance website (https://finance.yahoo.com/quote/BTC-USD/history/). The dataset has been slightly modified, as it has been provided alongside this chapter, and will be used throughout the rest of this book.
>
> The dataset can be downloaded from: https://packt.live/2Zgmm6r.

The following are the steps to complete this exercise:

1. Using your Terminal, navigate to the **Chapter02/Exercise2.01** directory. Activate the environment created in the previous chapter and execute the following command to start a Jupyter Notebook instance:

```
$ jupyter notebook
```

This should automatically open the Jupyter lab server in your browser. From there you can start a Jupyter Notebook.

You should see the following output or similar:

```
(base) MacBook-Pro-3:Exercise2.01 alexisrutherford$ jupyter-lab
[I 18:58:23.153 LabApp] The port 8888 is already in use, trying another port.
[I 18:58:23.155 LabApp] The port 8889 is already in use, trying another port.
[I 18:58:23.155 LabApp] The port 8890 is already in use, trying another port.
[I 18:58:23.156 LabApp] The port 8891 is already in use, trying another port.
[I 18:58:23.171 LabApp] JupyterLab extension loaded from /Users/alexisrutherford/opt/anaconda3/lib/python3.7/site-packages/jupyt
erlab
[I 18:58:23.171 LabApp] JupyterLab application directory is /Users/alexisrutherford/opt/anaconda3/share/jupyter/lab
[I 18:58:23.173 LabApp] Serving notebooks from local directory: /Users/alexisrutherford/The-Applied-TensorFlow-and-Keras-Worksho
p/Chapter02/Exercise2.01
[I 18:58:23.174 LabApp] The Jupyter Notebook is running at:
[I 18:58:23.174 LabApp] http://localhost:8892/?token=1d511cc6c12062a6124a4783466305d9b887bf6c2ee4cc9b
[I 18:58:23.174 LabApp]  or http://127.0.0.1:8892/?token=1d511cc6c12062a6124a4783466305d9b887bf6c2ee4cc9b
[I 18:58:23.174 LabApp] Use Control-C to stop this server and shut down all kernels (twice to skip confirmation).
[C 18:58:23.188 LabApp]

    To access the notebook, open this file in a browser:
        file:///Users/alexisrutherford/Library/Jupyter/runtime/nbserver-99125-open.html
    Or copy and paste one of these URLs:
        http://localhost:8892/?token=1d511cc6c12062a6124a4783466305d9b887bf6c2ee4cc9b
     or http://127.0.0.1:8892/?token=1d511cc6c12062a6124a4783466305d9b887bf6c2ee4cc9b
[I 18:58:26.275 LabApp] Build is up to date
```

Figure 2.9: Terminal image after starting a Jupyter lab instance

2. Select the **Exercise2.01_Exploring_Bitcoin_Dataset.ipynb** file. This is a Jupyter Notebook file that will open in a new browser tab. The application will automatically start a new Python interactive session for you:

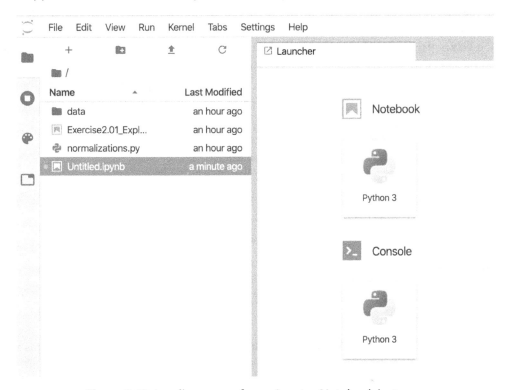

Figure 2.10: Landing page of your Jupyter Notebook instance

3. Click the Jupyter Notebook file:

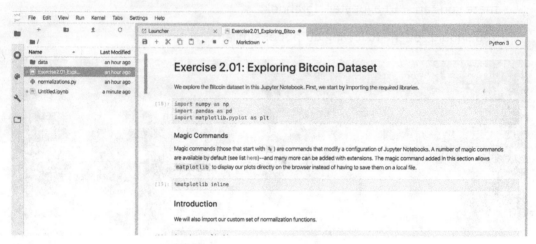

Figure 2.11: Image of Jupyter Notebook

4. Opening our Jupyter Notebook, consider the Bitcoin data made available with this chapter. The dataset **data/bitcoin_historical_prices.csv** (https://packt.live/2Zgmm6r) contains the details of Bitcoin prices since early 2013. It contains eight variables, two of which (**date** and **week**) describe a time period of the data. These can be used as indices—and six others (**open**, **high**, **low**, **close**, **volume**, and **market_capitalization**) can be used to understand changes in the price and value of Bitcoin over time:

Variable	Description
Date	Date of the observation
iso_week	Week number for a given year
Open	Open value for a single Bitcoin coin
High	Highest value achieved during a given day
low	Lowest value achieved during a given day
close	Value at the close of the transaction day
volume	The total volume of Bitcoin that was exchanged during that day
market_ capitalization	Market capitalization, which is explained by Market Cap = Price * Circulating Supply

Figure 2.12: Available variables (that is, columns) in the Bitcoin historical prices dataset

5. Using the open Jupyter Notebook instance, consider the time series of two of those variables: **close** and **volume**. Start with those time series to explore price fluctuation patterns, that is, how the price of Bitcoin varied at different times in the historical data.

6. Navigate to the open instance of the Jupyter Notebook, **Exercise2.01_Exploring_Bitcoin_Dataset.ipynb**. Now, execute all cells under the **Introduction** header. This will import the required libraries and import the dataset into memory:

Introduction

We will also import our custom set of normalization functions.

```
[3]: import normalizations
```

Let's load the dataset as a pandas `DataFrame`. This will make it easy to compute basic properties from the dataset and to clean any irregularities.

```
[4]: bitcoin = pd.read_csv('data/bitcoin_dataset.csv', date_parser=['date'])
```

```
[5]: bitcoin.head()
```

[5]:		date	open	high	low	close	volume	iso_week
	0	2014-09-27	403.556000	406.622986	397.372009	399.519989	15029300	2014-39
	1	2014-09-28	399.471008	401.016998	374.332001	377.181000	23613300	2014-39
	2	2014-09-29	376.928009	385.210999	372.239990	375.467010	32497700	2014-40
	3	2014-09-30	376.088013	390.976990	373.442993	386.944000	34707300	2014-40
	4	2014-10-01	387.427002	391.378998	380.779999	383.614990	26229400	2014-40

Our dataset contains 7 variables (i.e. columns). Here's what each one of them represents:

- `date` : date of the observation.
- `open` : open value of a single Bitcoin coin.
- `high` : highest value achieved during a given day period.
- `low` : lowest value achieved during a given day period.

Figure 2.13: Output from the first cells of the notebook time-series plot of the closing price for Bitcoin from the close variable

7. After the dataset has been imported into memory, move to the **Exploration** section. You will find a snippet of code that generates a time series plot for the **close** variable:

```
bitcoin.set_index('date')['close'].plot(linewidth=2, \
                                         figsize=(14, 4),\
                                         color='#d35400')

#plt.plot(bitcoin['date'], bitcoin['close'])
```

The output looks like:

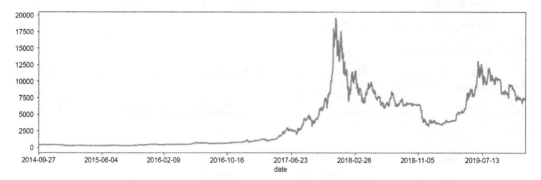

Figure 2.14: Time series plot of the closing price for Bitcoin from the close variable

8. Reproduce this plot but using the **volume** variable in a new cell below this one. You will have most certainly noticed that price variables surge in 2017 and then the downfall starts:

```
bitcoin.set_index('date')['volume'].plot(linewidth=2, \
                                          figsize=(14, 4), \
                                          color='#d35400')
```

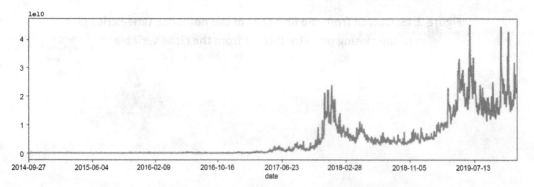

Figure 2.15: The total daily volume of Bitcoin coins

Figure 2.15 shows that since 2017, Bitcoin transactions have significantly increased in the market. The total daily volume varies much more than daily closing prices.

9. Execute the remaining cells in the Exploration section to explore the range from 2017 to 2018.

Fluctuations in Bitcoin prices have been increasingly commonplace in recent years. While those periods could be used by a neural network to understand certain patterns, we will be excluding older observations, given that we are interested in predicting future prices for not-too-distant periods. Filter the data after 2016 only. Navigate to the **Preparing Dataset for a Model** section. Use the pandas API to filter the data. Pandas provides an intuitive API for performing this operation.

10. Extract recent data and save it into a variable:

```
bitcoin_recent = bitcoin[bitcoin['date'] >= '2016-01-04']
```

The **bitcoin_recent** variable now has a copy of our original Bitcoin dataset, but filtered to the observations that are newer or equal to January 4, 2016.

Normalize the data using the point-relative normalization technique described in the *Data Normalization* section in the Jupyter Notebook. You will only normalize two variables—**close** and **volume**—because those are the variables that we are working to predict.

11. Run the next cell in the notebook to ensure that we only keep the close and volume variables.

In the same directory containing this chapter, we have placed a script called **normalizations.py**. That script contains the three normalization techniques described in this chapter. We import that script into our Jupyter Notebook and apply the functions to our series.

12. Navigate to the **Preparing Dataset for a Model** section. Now, use the **iso_week** variable to group daily observations from a given week using the pandas **groupby()** method. We can now apply the normalization function, **normalizations.point_relative_normalization()**, directly to the series within that week. We can store the normalization output as a new variable in the same pandas DataFrame using the following code:

```
bitcoin_recent['close_point_relative_normalization'] = \
bitcoin_recent.groupby('iso_week')['close']\
.apply(lambda x: normalizations.point_relative_normalization(x))
```

13. The **close_point_relative_normalization** variable now contains the normalized data for the **close** variable:

```
bitcoin_recent.set_index('date')\
['close_point_relative_normalization'].plot(linewidth=2, \
                                            figsize=(14,4), \
                                            color='#d35400')
```

This will result in the following output:

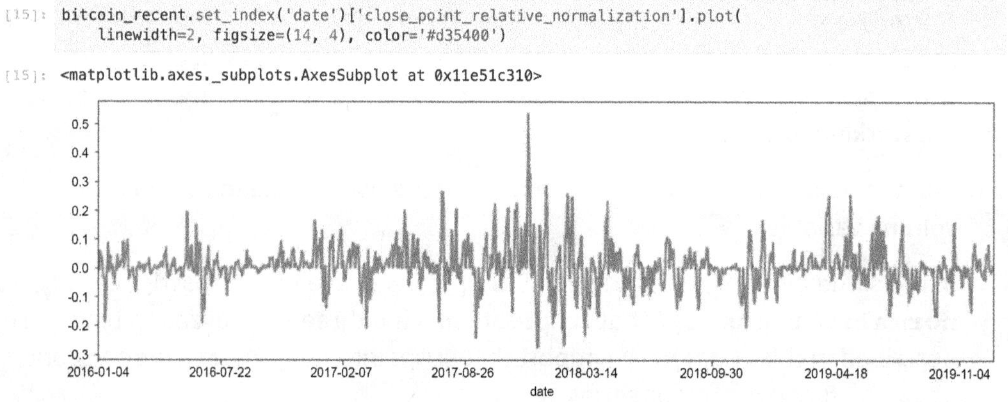

```
[15]:  bitcoin_recent.set_index('date')['close_point_relative_normalization'].plot(
           linewidth=2, figsize=(14, 4), color='#d35400')
```

```
[15]:  <matplotlib.axes._subplots.AxesSubplot at 0x11e51c310>
```

Figure 2.16: Image of the Jupyter Notebook focusing on the section where the normalization function is applied

14. Do the same with the **volume** variable (**volume_point_relative_normalization**). The normalized **close** variable contains an interesting variance pattern every week. We will be using that variable to train our LSTM model:

```
bitcoin_recent.set_index('date')\
                        ['volume_point_relative_normalization'].\
                        plot(linewidth=2, \
                        figsize=(14,4), \
                        color='#d35400')
```

Your output should be as follows.

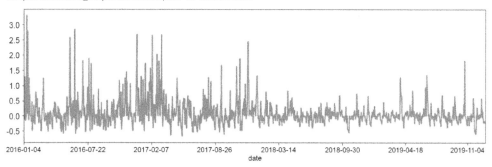

```
[18]: <matplotlib.axes._subplots.AxesSubplot at 0x12a37bf90>
```

Figure 2.17: Plot that displays the series from the normalized variable

15. In order to evaluate how well the model performs, you need to test its accuracy versus some other data. Do this by creating two datasets: a training set and a test set. You will use 90 percent of the dataset to train the LSTM model and 10 percent to evaluate its performance. Given that the data is continuous and in the form of a time series, use the last 10 percent of available weeks as a test set and the first 90 percent as a training set:

```
boundary = int(0.9 * bitcoin_recent['iso_week'].nunique())
train_set_weeks = bitcoin_recent['iso_week'].unique()[0:boundary]
test_set_weeks = bitcoin_recent[~bitcoin_recent['iso_week']\
                .isin(train_set_weeks)]['iso_week'].unique()

test_set_weeks
train_set_weeks
```

This will display the following output:

```
[21]: test_set_weeks
```

```
[21]: array(['2019-32', '2019-33', '2019-34', '2019-35', '2019-36', '2019-37',
             '2019-38', '2019-39', '2019-40', '2019-41', '2019-42', '2019-43',
             '2019-44', '2019-45', '2019-46', '2019-47', '2019-48', '2019-49',
             '2019-50', '2019-51', '2019-52'], dtype=object)
```

Figure 2.18: Output of the test set weeks

```
[20]: train_set_weeks
```

```
[20]: array(['2016-1', '2016-2', '2016-3', '2016-4', '2016-5', '2016-6',
             '2016-7', '2016-8', '2016-9', '2016-10', '2016-11', '2016-12',
             '2016-13', '2016-14', '2016-15', '2016-16', '2016-17', '2016-18',
             '2016-19', '2016-20', '2016-21', '2016-22', '2016-23', '2016-24',
             '2016-25', '2016-26', '2016-27', '2016-28', '2016-29', '2016-30',
             '2016-31', '2016-32', '2016-33', '2016-34', '2016-35', '2016-36',
             '2016-37', '2016-38', '2016-39', '2016-40', '2016-41', '2016-42',
             '2016-43', '2016-44', '2016-45', '2016-46', '2016-47', '2016-48',
             '2016-49', '2016-50', '2016-51', '2016-52', '2017-1', '2017-2',
             '2017-3', '2017-4', '2017-5', '2017-6', '2017-7', '2017-8',
             '2017-9', '2017-10', '2017-11', '2017-12', '2017-13', '2017-14',
             '2017-15', '2017-16', '2017-17', '2017-18', '2017-19', '2017-20',
             '2017-21', '2017-22', '2017-23', '2017-24', '2017-25', '2017-26',
             '2017-27', '2017-28', '2017-29', '2017-30', '2017-31', '2017-32',
             '2017-33', '2017-34', '2017-35', '2017-36', '2017-37', '2017-38',
             '2017-39', '2017-40', '2017-41', '2017-42', '2017-43', '2017-44',
             '2017-45', '2017-46', '2017-47', '2017-48', '2017-49', '2017-50',
             '2017-51', '2017-52', '2018-1', '2018-2', '2018-3', '2018-4',
             '2018-5', '2018-6', '2018-7', '2018-8', '2018-9', '2018-10',
             '2018-11', '2018-12', '2018-13', '2018-14', '2018-15', '2018-16',
             '2018-17', '2018-18', '2018-19', '2018-20', '2018-21', '2018-22',
             '2018-23', '2018-24', '2018-25', '2018-26', '2018-27', '2018-28',
             '2018-29', '2018-30', '2018-31', '2018-32', '2018-33', '2018-34',
             '2018-35', '2018-36', '2018-37', '2018-38', '2018-39', '2018-40',
             '2018-41', '2018-42', '2018-43', '2018-44', '2018-45', '2018-46',
             '2018-47', '2018-48', '2018-49', '2018-50', '2018-51', '2018-52',
             '2019-1', '2019-2', '2019-3', '2019-4', '2019-5', '2019-6',
             '2019-7', '2019-8', '2019-9', '2019-10', '2019-11', '2019-12',
             '2019-13', '2019-14', '2019-15', '2019-16', '2019-17', '2019-18',
             '2019-19', '2019-20', '2019-21', '2019-22', '2019-23', '2019-24',
             '2019-25', '2019-26', '2019-27', '2019-28', '2019-29', '2019-30',
             '2019-31'], dtype=object)
```

Figure 2.19: Using weeks to create a training set

16. Create the separate datasets for each operation:

```
train_dataset = bitcoin_recent[bitcoin_recent['iso_week']\
                                        .isin(train_set_weeks)]
test_dataset = bitcoin_recent[bitcoin_recent['iso_week'].\
                                        isin(test_set_weeks)]
```

17. Finally, navigate to the **Storing Output** section and save the filtered variable to disk, as follows:

```
test_dataset.to_csv('data/test_dataset.csv', index=False)
train_dataset.to_csv('data/train_dataset.csv', index=False)
bitcoin_recent.to_csv('data/bitcoin_recent.csv', index=False)
```

> **NOTE**
>
> To access the source code for this specific section, please refer to https://packt.live/3ehbgCi.
>
> You can also run this example online at https://packt.live/2ZdGq9s. You must execute the entire Notebook in order to get the desired result.

In this exercise, we explored the Bitcoin dataset and prepared it for a deep learning model.

We learned that in 2017, the price of Bitcoin skyrocketed. This phenomenon took a long time to take place and may have been influenced by a number of external factors that this data alone doesn't explain (for instance, the emergence of other cryptocurrencies). After the great surge of 2017, we saw a great fall in the value of Bitcoin in 2018.

We also used the point-relative normalization technique to process the Bitcoin dataset in weekly chunks. We do this to train an LSTM network to learn the weekly patterns of Bitcoin price changes so that it can predict a full week into the future.

However, Bitcoin statistics show significant fluctuations on a weekly basis. Can we predict the price of Bitcoin in the future? What will the price be seven days from now? We will build a deep learning model to explore these questions in our next section using Keras.

USING KERAS AS A TENSORFLOW INTERFACE

We are using Keras because it simplifies the TensorFlow interface into general abstractions and, in TensorFlow 2.0, this is the default API in this version. In the backend, the computations are still performed in TensorFlow, but we spend less time worrying about individual components, such as variables and operations, and spend more time building the network as a computational unit. Keras makes it easy to experiment with different architectures and hyperparameters, moving more quickly toward a performant solution.

As of TensorFlow 2.0.0, Keras is now officially distributed with TensorFlow as **tf.keras**. This suggests that Keras is now tightly integrated with TensorFlow and will likely continue to be developed as an open source tool for a long period of time. Components are an integral part when building models. Let's deep dive into this concept now.

MODEL COMPONENTS

As we saw in *Chapter 1, Introduction to Neural Networks and Deep Learning*, LSTM networks also have input, hidden, and output layers. Each hidden layer has an activation function that evaluates that layer's associated weights and biases. As expected, the network moves data sequentially from one layer to another and evaluates the results by the output at every iteration (that is, an epoch).

Keras provides intuitive classes that represent each one of the components listed in the following table:

Component	Keras Class
High-level abstraction of a complete sequential neural network.	`keras.models.Sequential()`
Dense, fully connected layer.	`keras.layers.core.Dense()`
Activation function.	`keras.layers.core.Activation()`
LSTM recurrent neural network. This class contains components that are exclusive to this architecture, most of which are abstracted by Keras.	`keras.layers.recurrent.LSTM(`

Figure 2.20: Description of key components from the Keras API

We will be using these components to build a deep learning model.

Keras' **keras.models.Sequential()** component represents a whole sequential neural network. This Python class can be instantiated on its own and have other components added to it subsequently.

We are interested in building an LSTM network because those networks perform well with sequential data—and a time series is a kind of sequential data. Using Keras, the complete LSTM network would be implemented as follows:

```
from tensorflow.keras.models import Sequential
from tensorflow.keras.layers import LSTM
from tensorflow.keras.layers import Dense, Activation
model = Sequential()
model.add(LSTM(units=number_of_periods, \
                input_shape=(period_length, number_of_periods) \
                return_sequences=False), stateful=True)
model.add(Dense(units=period_length)) \
        model.add(Activation("linear"))
model.compile(loss="mse", optimizer="rmsprop")
```

This implementation will be further optimized in *Chapter 3, Real-World Deep Learning: Evaluating the Bitcoin Model*.

Keras abstraction allows you to focus on the key elements that make a deep learning system more performant: determining the right sequence of components, how many layers and nodes to include, and which activation function to use. All of these choices are determined by either the order in which components are added to the instantiated **keras.models.Sequential()** class or by parameters passed to each component instantiation (that is, **Activation("linear")**). The final **model.compile()** step builds the neural network using TensorFlow components.

After the network is built, we train our network using the **model.fit()** method. This will yield a trained model that can be used to make predictions:

```
model.fit(X_train, Y_train,
            batch_size=32, epochs=epochs)
```

The **X_train** and **Y_train** variables are, respectively, a set used for training and a smaller set used for evaluating the loss function (that is, testing how well the network predicts data). Finally, we can make predictions using the **model.predict()** method:

```
model.predict(x=X_train)
```

The preceding steps cover the Keras paradigm for working with neural networks. Despite the fact that different architectures can be dealt with in very different ways, Keras simplifies the interface for working with different architectures by using three components – **Network Architecture**, **Fit**, and **Predict**:

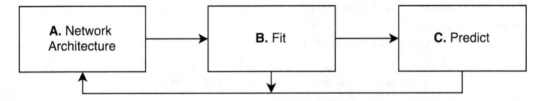

Figure 2.21: The Keras neural network paradigm

The Keras neural network diagram comprises the following three steps:

• A neural network architecture

• Training a neural network (or **Fit**)

• Making predictions

Keras allows much greater control within each of these steps. However, its focus is to make it as easy as possible for users to create neural networks in as little time as possible. That means that we can start with a simple model, and then add complexity to each one of the preceding steps to make that initial model perform better.

We will take advantage of that paradigm during our upcoming exercise and chapters. In the next exercise, we will create the simplest LSTM network possible. Then, in *Chapter 3, Real-World Deep Learning: Evaluating the Bitcoin Model*, we will continuously evaluate and alter that network to make it more robust and performant.

EXERCISE 2.02: CREATING A TENSORFLOW MODEL USING KERAS

In this notebook, we design and compile a deep learning model using Keras as an interface to TensorFlow. We will continue to modify this model in our next chapters and exercises by experimenting with different optimization techniques. However, the essential components of the model are designed entirely in this notebook:

1. Open a new Jupyter Notebook and import the following libraries:

```
import warnings
warnings.filterwarnings("ignore", category=DeprecationWarning)
import tensorflow as tf
from tensorflow import keras
from tensorflow.keras.models import Sequential
from tensorflow.keras.layers import LSTM
from tensorflow.keras.layers import Dense, Activation
```

2. Our dataset contains daily observations and each observation influences a future observation. Also, we are interested in predicting a week—that is, 7 days—of Bitcoin prices in the future:

```
period_length = 7
number_of_periods = 208 - 21 - 1
number_of_periods
```

We have calculated **number_of_observations** based on available weeks in our dataset. Given that we will be using last week to test the LSTM network on every epoch, we will use 208 – 21 – 1. You'll get:

```
186
```

3. Build the LSTM model using Keras. We have the batch size as one because we are passing the whole data in a single iteration. If data is big, then we can pass the data with multiple batches, That's why we used batch_input_shape:

```
def build_model(period_length, number_of_periods, batch_size=1):
    model = Sequential()
    model.add(LSTM(units=period_length,\
                    batch_input_shape=(batch_size, \
                                        number_of_periods, \
                                        period_length),\
                    input_shape=(number_of_periods, \
                                    period_length),\
                    return_sequences=False, stateful=False))

    model.add(Dense(units=period_length))
    model.add(Activation("linear"))

    model.compile(loss="mse", optimizer="rmsprop")

    return model
```

This should return a compiled Keras model that can be trained and stored in disk.

4. Let's store the model on the output of the model to a disk:

```
model = build_model(period_length=period_length, \
                    number_of_periods=number_of_periods)
model.save('bitcoin_lstm_v0.h5')
```

Note that the **bitcoin_lstm_v0.h5** model hasn't been trained yet. When saving a model without prior training, you effectively only save the architecture of the model. That same model can later be loaded by using Keras' **load_model()** function, as follows:

```
model = keras.models.load_model('bitcoin_lstm_v0.h5')
```

> **NOTE**
>
> To access the source code for this specific section, please refer to
> https://packt.live/38KQI3Y.
>
> You can also run this example online at https://packt.live/3fhEL89.
> You must execute the entire Notebook in order to get the desired result.

This concludes the creation of our Keras model, which we can now use to make predictions.

> **NOTE**
>
> You may encounter the following warning when loading the Keras library:
>
> **Using TensorFlow backend**
>
> Keras can be configured to use another backend instead of TensorFlow (that is, Theano). In order to avoid this message, you can create a file called **keras.json** and configure its backend there. The correct configuration of that file depends on your system. Hence, it is recommended that you visit Keras' official documentation on the topic at https://keras.io/backend/.

In this section, we have learned how to build a deep learning model using Keras—an interface for TensorFlow. We studied core components of Keras and used those components to build the first version of our Bitcoin price-predicting system based on an LSTM model.

In our next section, we will discuss how to put all the components from this chapter together into a (nearly complete) deep learning system. That system will yield our very first predictions, serving as a starting point for future improvements.

FROM DATA PREPARATION TO MODELING

This section focuses on the implementation aspects of a deep learning system. We will use the Bitcoin data from the *Choosing the Right Model Architecture* section, and the Keras knowledge from the preceding section, *Using Keras as a TensorFlow Interface*, to put both of these components together. This section concludes the chapter by building a system that reads data from a disk and feeds it into a model as a single piece of software.

TRAINING A NEURAL NETWORK

Neural networks can take long periods of time to train. Many factors affect how long that process may take. Among them, three factors are commonly considered the most important:

- The network's architecture

- How many layers and neurons the network has

- How much data there is to be used in the training process

Other factors may also greatly impact how long a network takes to train, but most of the optimization that a neural network can have when addressing a business problem comes from exploring those three.

We will be using the normalized data from our previous section. Recall that we have stored the training data in a file called **train_dataset.csv**.

> **NOTE:**
>
> You can download the training data by visiting this link:
> https://packt.live/2Zgmm6r.

We will load that dataset into memory using the **pandas** library for easy exploration:

```
import pandas as pd
train = pd.read_csv('data/train_dataset.csv')
```

> **NOTE**
>
> Make sure you change the path (highlighted) based on where you have
> downloaded or saved the CSV file.

You will see the output in a tabular form as follows:

	date	open	high	low	close	volume	iso_week
0	2014-09-27	403.556000	406.622986	397.372009	399.519989	15029300	2014-39
1	2014-09-28	399.471008	401.016998	374.332001	377.181000	23613300	2014-39
2	2014-09-29	376.928009	385.210999	372.239990	375.467010	32497700	2014-40
3	2014-09-30	376.088013	390.976990	373.442993	386.944000	34707300	2014-40
4	2014-10-01	387.427002	391.378998	380.779999	383.614990	26229400	2014-40

Figure 2.22: Table showing the first five rows of the training dataset

We will be using the series from the **close_point_relative_normalization** variable, which is a normalized series of the Bitcoin closing prices—from the **close** variable—since the beginning of 2016.

The **close_point_relative_normalization** variable has been normalized on a weekly basis. Each observation from the week's period is made relative to the difference from the closing prices on the first day of the period. This normalization step is important and will help our network train faster:

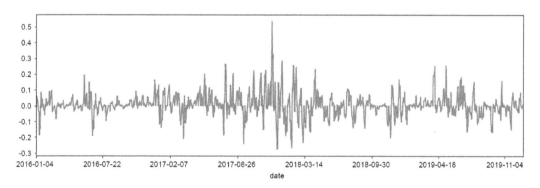

Figure 2.23: Plot that displays the series from the normalized variable.

This variable will be used to train our LSTM model.

RESHAPING TIME SERIES DATA

Neural networks typically work with vectors and tensors—both mathematical objects that organize data in a number of dimensions. Each neural network implemented in Keras will have either a vector or a tensor that is organized according to a specification as input.

At first, understanding how to reshape the data into the format expected by a given layer can be confusing. To avoid confusion, it is advisable to start with a network with as few components as possible, and then add components gradually. Keras' official documentation (under the **Layers** section) is essential for learning about the requirements for each kind of layer.

> **NOTE**
>
> The Keras official documentation is available at https://keras.io/layers/core/. That link takes you directly to the **Layers** section.

NumPy is a popular Python library used for performing numerical computations. It is used by the deep learning community to manipulate vectors and tensors and prepare them for deep learning systems.

In particular, the **numpy.reshape()** method is very important when adapting data for deep learning models. That model allows for the manipulation of NumPy arrays, which are Python objects analogous to vectors and tensors.

We'll now organize the prices from the **close_point_relative_normalization** variable using the weeks after 2016. We will create distinct groups containing 7 observations each (one for each day of the week) for a total of 208 complete weeks. We do that because we are interested in predicting the prices of a week's worth of trading.

> **NOTE**
>
> We use the ISO standard to determine the beginning and the end of a week. Other kinds of organizations are entirely possible. This one is simple and intuitive to follow, but there is room for improvement.

LSTM networks work with three-dimensional tensors. Each one of those dimensions represents an important property for the network. These dimensions are as follows:

- **Period length**: The period length, that is, how many observations there are for a period

- **Number of periods**: How many periods are available in the dataset

- **Number of features**: The number of features available in the dataset

Our data from the **close_point_relative_normalization** variable is currently a one-dimensional vector. We need to reshape it to match those three dimensions.

We will be using a period of a week. So, our period length is 7 days (period length = 7). We have 208 complete weeks available in our data. We will be using the very last of those weeks to test our model against during its training period. That leaves us with 187 distinct weeks. Finally, we will be using a single feature in this network (number of features = 1); we will include more features in future versions.

To reshape the data to match those dimensions, we will be using a combination of base Python properties and the **reshape()** method from the **numpy** library. First, we create the 186 distinct week groups with 7 days, each using pure Python:

```
group_size = 7
samples = list()
for i in range(0, len(data), group_size):
sample = list(data[i:i + group_size])
        if len(sample) == group_size:samples\
                    .append(np.array(sample)\
                    .reshape(group_size, 1).tolist())

data = np.array(samples)
```

This piece of code creates distinct week groups. The resulting variable data is a variable that contains all the right dimensions.

> **NOTE**
>
> Each Keras layer will expect its input to be organized in specific ways. However, Keras will reshape data accordingly, in most cases. Always refer to the Keras documentation on layers (https://keras.io/layers/core/) before adding a new layer.

The Keras LSTM layer expects these dimensions to be organized in a specific order: the number of features, the number of observations, and the period length. Reshape the dataset to match that format:

```
X_train = data[:-1,:].reshape(1, 186, 7)
Y_validation = data[-1].reshape(1, 7)
```

The preceding snippet also selects the very last week of our set as a validation set (via **data[-1]**). We will be attempting to predict the very last week in our dataset by using the preceding 76 weeks. The next step is to use those variables to fit our model:

```
model.fit(x=X_train, y=Y_validation, epochs=100)
```

LSTMs are computationally expensive models. They may take up to 5 minutes to train with our dataset on a modern computer. Most of that time is spent at the beginning of the computation when the algorithm creates the full computation graph. The process gains speed after it starts training:

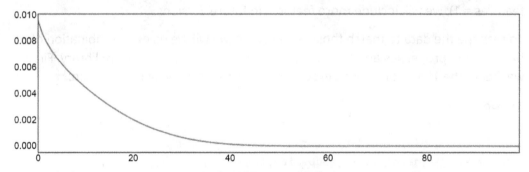

Figure 2.24: Graph that shows the results of the loss function evaluated at each epoch

> **NOTE**
>
> This compares what the model predicted at each epoch, and then compares that with the real data using a technique called mean-squared error. This plot shows those results.

At a glance, our network seems to perform very well; it starts with a very small error rate that continuously decreases. Now that we have lowered the error rate, let's move on to make some predictions.

MAKING PREDICTIONS

After our network has been trained, we can proceed to make predictions. We will be making predictions for a future week beyond our time period.

Once we have trained our model with the **model.fit()** method, making predictions is simple:

```
model.predict(x=X_train)
```

We use the same data for making predictions as the data used for training (the **X_train** variable). If we have more data available, we can use that instead—given that we reshape it to the format the LSTM requires.

OVERFITTING

When a neural network overfits a validation set, this means that it learns patterns present in the training set but is unable to generalize it to unseen data (for instance, the test set). In the next chapter, we will learn how to avoid overfitting and create a system for both evaluating our network and increasing its performance:

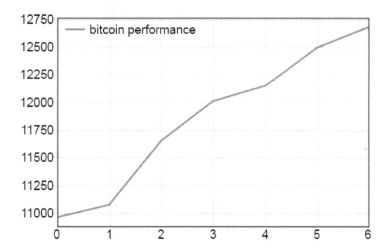

Figure 2.25: Graph showing the weekly performance of Bitcoin

In the plot shown above, the horizontal axis represents the week number and the vertical axis represents the predicted performance of Bitcoin. Now that we have explored the data, prepared a model, and learned how to make predictions, let's put this knowledge into practice.

ACTIVITY 2.01: ASSEMBLING A DEEP LEARNING SYSTEM

In this activity, we'll bring together all the essential pieces for building a basic deep learning system – the data, a model, and predictions:

1. Start a Jupyter Notebook.

2. Load the training dataset into memory.

3. Inspect the training set to see whether it is in the form period length, number of periods, or number of features.

4. Convert the training set if it is not in the required format.

5. Load the previously trained model.

6. Train the model using your training dataset.

7. Make a prediction on the training set.

8. Denormalize the values and save the model.

The final output will look as follows with the horizontal axis representing the number of days and the vertical axis represents the price of Bitcoin:

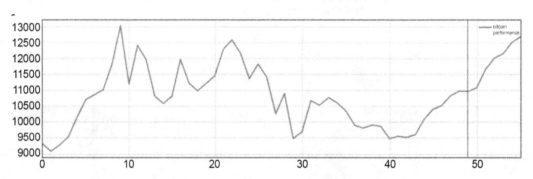

Figure 2.26: Expected output

> **NOTE**
>
> The solution to this activity can be found on page 136.

SUMMARY

In this chapter, we have assembled a complete deep learning system, from data to prediction. The model created in this activity requires a number of improvements before it can be considered useful. However, it serves as a great starting point from which we will continuously improve.

The next chapter will explore techniques for measuring the performance of our model and will continue to make modifications until we reach a model that is both useful and robust.

3

REAL-WORLD DEEP LEARNING: EVALUATING THE BITCOIN MODEL

OVERVIEW

This chapter focuses on how to evaluate a neural network model. We'll modify the network's hyperparameters to improve its performance. However, before altering any parameters, we need to measure how the model performs. By the end of this chapter, you will be able to evaluate a model using different functions and techniques. You will also learn about hypermeter optimization by implementing functions and regularization strategies.

INTRODUCTION

In the previous chapter, you trained your model. But how will you check its performance and whether it is performing well or not? Let's find out by evaluating a model. In machine learning, it is common to define two distinct terms: **parameter** and **hyperparameter**. Parameters are properties that affect how a model makes predictions from data, say from a particular dataset. Hyperparameters refer to how a model learns from data. Parameters can be learned from the data and modified dynamically. Hyperparameters, on the other hand, are higher-level properties defined before the training begins and are not typically learned from data. In this chapter, you will learn about these factors in detail and understand how to use them with different evaluation techniques to improve the performance of a model.

> **NOTE**
>
> For a more detailed overview of machine learning, refer to *Python Machine Learning, Sebastian Raschka and Vahid Mirjalili, Packt Publishing, 2017).*

PROBLEM CATEGORIES

Generally, there are two categories of problems that can be solved by neural networks: **classification** and **regression**. Classification problems concern the prediction of the right categories from data—for instance, whether the temperature is hot or cold. Regression problems are about the prediction of values in a continuous scalar—for instance, what the actual temperature value is.

The problems in these two categories are characterized by the following properties:

* **Classification**: These are problems that are characterized by categories. The categories can be different, or not. They can also be about a binary problem, where the outcome can either be **yes** or **no**. However, they must be clearly assigned to each data element. An example of a classification problem would be to assign either the **car** or **not a car** labels to an image using a convolutional neural network. The MNIST example we explored in *Chapter 1, Introduction to Neural Networks and Deep Learning*, is another example of a classification problem.

- **Regression**: These are problems that are characterized by a continuous variable (that is, a scalar). They are measured in terms of ranges, and their evaluations consider how close to the real values the network is. An example is a time-series classification problem in which a recurrent neural network is used to predict the future temperature values. The Bitcoin price-prediction problem is another example of a regression problem.

While the overall structure of how to evaluate these models is the same for both of these problem categories, we employ different techniques to evaluate how models perform. In the next section, we'll explore these techniques for either classification or regression problems.

LOSS FUNCTIONS, ACCURACY, AND ERROR RATES

Neural networks utilize functions that measure how the networks perform compared to a **validation set**—that is, a part of the data that is kept separate to be used as part of the training process. These functions are called **loss functions**.

Loss functions evaluate how wrong a neural network's predictions are. Then, they propagate those errors back and make adjustments to the network, modifying how individual neurons are activated. Loss functions are key components of neural networks, and choosing the right loss function can have a significant impact on how the network performs. Errors are propagated through a process called **backpropagation**, which is a technique for propagating the errors that are returned by the loss function to each neuron in a neural network. Propagated errors affect how neurons activate, and ultimately, how they influence the output of that network.

Many neural network packages, including Keras, use this technique by default.

> **NOTE**
>
> For more information about the mathematics of backpropagation, please refer to *Deep Learning* by *Ian Goodfellow et. al., MIT Press, 2016*.

We use different loss functions for regression and classification problems. For classification problems, we use accuracy functions (that is, the proportion of the number of times the predictions were correct to the number of times predictions were made). For example, if you predict a toss of a coin that will result in m times as heads when you toss it n times and your prediction is correct, then the accuracy will be calculated as m/n. For regression problems, we use error rates (that is, how close the predicted values were to the observed ones).

Here's a summary of common loss functions that can be utilized, alongside their common applications:

Problem Type	Loss Function	Problem	Example
Regression	Mean Squared Error (MSE)	Predicting a continuous function. That is, predicting a value within a range of values.	Predicting the temperature in the future using temperature measurements from the past.
Regression	Root Mean Squared Error (RMSE)	Same as preceding, but deals with negative values. RMSE typically provides more interpretable results.	Same as preceding.
Regression	Mean Absolute Percentage Error (MAPE)	Predicting continuous functions. Has better performance when working with de-normalized ranges.	Predicting the sales for a product using the product properties (for example, price, type, target audience, market conditions).
Classification	Binary Cross-entropy	Classification between two categories or between two values (that is, true or false).	Predicting if the visitor of a website is male or female based on their browser activity.
Classification	Categorical Cross-entropy	Classification between many categories from a known set of categories.	Predicting the nationality of a speaker based on their accent when speaking a sentence in English.

Figure 3.1: Common loss functions used for classification and regression problems

For regression problems, the MSE function is the most common choice, while for classification problems, binary cross-entropy (for binary category problems) and categorical cross-entropy (for multi-category problems) are common choices. It is advised to start with these loss functions, then experiment with other functions as you evolve your neural network, aiming to gain performance.

The network we developed in *Chapter 2, Real-World Deep Learning: Predicting the Price of Bitcoin*, uses MSE as its loss function. In the next section, we'll explore how that function performs as the network trains.

DIFFERENT LOSS FUNCTIONS, SAME ARCHITECTURE

Before moving ahead to the next section, let's explore, in practical terms, how these problems are different in the context of neural networks.

The *TensorFlow Playground* (https://playground.tensorflow.org/) application has been made available by the TensorFlow team to help us understand how neural networks work. Here, we can see a neural network represented with its layers: input (on the left), hidden layers (in the middle), and output (on the right).

> **NOTE:**
>
> These images can be viewed in the repository on GitHub at: https://packt.live/2Cl1t0H.

We can also choose different sample datasets to experiment with on the far left-hand side. And, finally, on the far right-hand side, we can see the output of the network:

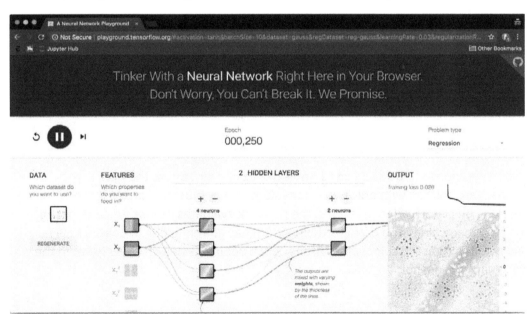

Figure 3.2: TensorFlow Playground web application

Take the parameters for a neural network shown in this visualization to gain an idea of how each parameter affects the model's results. This application helps us explore the different problem categories we discussed in the previous section. When we choose **Regression** (upper right-hand corner), the colors of the dots are colored in a range of color values between orange and blue:

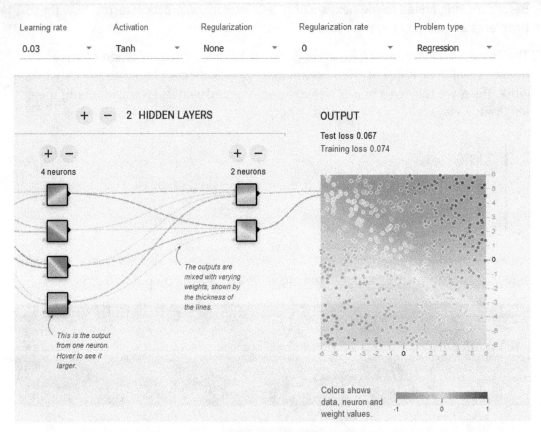

Figure 3.3: Regression problem example in TensorFlow Playground

When we choose **Classification** as the **Problem type**, the dots in the dataset are colored with only two color values: either blue or orange. When working on classification problems, the network evaluates its loss function based on how many blues and oranges the network has gotten wrong. It checks how far away to the right the color values are for each dot in the network, as shown in the following screenshot:

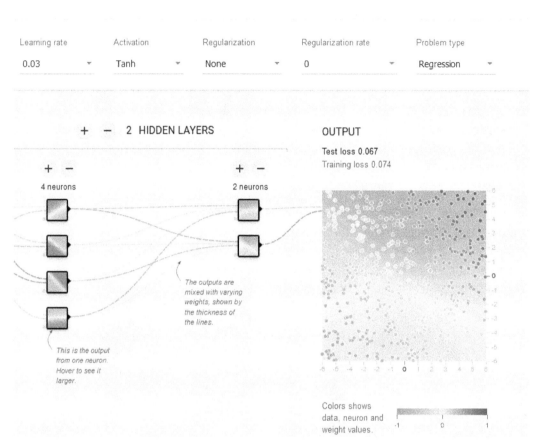

Figure 3.4: Details of the TensorFlow Playground application. Different colors are assigned to different classes in classification problems

After clicking on the play button, we notice that the numbers in the `Training loss` area keep going down as the network continuously trains. The numbers are very similar in each problem category because the loss functions play the same role in both neural networks.

However, the actual loss function that's used for each category is different and is chosen depending on the problem type.

USING TENSORBOARD

Evaluating neural networks is where TensorBoard excels. As we explained in *Chapter 1, Introduction to Neural Networks and Deep Learning*, TensorBoard is a suite of visualization tools that's shipped with TensorFlow. Among other things, we can explore the results of loss function evaluations after each epoch. A great feature of TensorBoard is that we can organize the results of each run separately and compare the resulting loss function metrics for each run. We can then decide on which hyperparameters to tune and have a general sense of how the network is performing. The best part is that this is all done in real time.

In order to use TensorBoard with our model, we will use Keras' **callback** function. We do this by importing the TensorBoard **callback** and passing it to our model when calling its **fit()** function. The following code shows an example of how this would be implemented in the Bitcoin model we created in the *Chapter 2, Real-World Deep Learning: Predicting the Price of Bitcoin*:

```
from tensorflow.keras.callbacks import TensorBoard
model_name = 'bitcoin_lstm_v0_run_0'
tensorboard = TensorBoard(log_dir='logs\\{}'.format(model_name)) \
                    model.fit(x=X_train, y=Y_validate, \
                    batch_size=1, epochs=100, verbose=0, \
                    callbacks=[tensorboard])
```

Keras **callback** functions are called at the end of each epoch run. In this case, Keras calls the TensorBoard **callback** to store the results from each run on the disk. There are many other useful **callback** functions available, and you can create custom ones using the Keras API.

> **NOTE**
>
> Please refer to the Keras callback documentation (https://keras.io/callbacks/) for more information.

After implementing the TensorBoard callback, the loss function metrics are now available in the TensorBoard interface. You can now run a TensorBoard process (with **tensorboard --logdir=./logs**) and leave it running while you train your network with **fit()**.

The main graphic to evaluate is typically called loss. We can add more metrics by passing known metrics to the metrics parameter in the **fit()** function. These will then be available for visualization in TensorBoard, but will not be used to adjust the network weights. The interactive graphics will continue to update in real time, which allows you to understand what is happening on every epoch. In the following screenshot, you can see a TensorBoard instance showing loss function results, alongside other metrics that have been added to the metrics parameter:

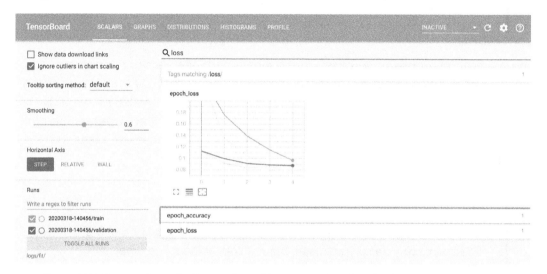

Figure 3.5: Screenshot of a TensorBoard instance showing the loss function results

In the next section, we will talk more about how to implement the different metrics we discussed in this section.

IMPLEMENTING MODEL EVALUATION METRICS

In both regression and classification problems, we split the input dataset into three other datasets: train, validation, and test. Both the train and the validation sets are used to train the network. The train set is used by the network as input, while the validation set is used by the loss function to compare the output of the neural network to the real data and compute how wrong the predictions are. Finally, the test set is used after the network has been trained to measure how the network can perform on data it has never seen before.

> **NOTE**
>
> There isn't a clear rule for determining how the train, validation, and test datasets must be divided. It is a common approach to divide the original dataset into 80 percent train and 20 percent test, then to further divide the train dataset into 80 percent train and 20 percent validation. For more information about this problem, please refer to *Python Machine Learning*, by *Sebastian Raschka and Vahid Mirjalili (Packt Publishing, 2017)*.

In classification problems, you pass both the data and the labels to the neural network as related but distinct data. The network then learns how the data is related to each label. In regression problems, instead of passing data and labels, we pass the variable of interest as one parameter and the variables that are used for learning patterns as another. Keras provides an interface for both of those use cases with the **fit()** method.

> **NOTE**
>
> The **fit()** method can use either the **validation_split** or the **validation_data** parameter, but not both at the same time.

See the following snippet to understand how to use the **validation_split** and **validation_data** parameters:

```
model.fit(x=X_train, y=Y_ train, \
          batch_size=1, epochs=100, verbose=0, \
          callbacks=[tensorboard], validation_split=0.1, \
          validation_data=(X_validation, Y_validation))
```

X_train: features from training set

Y_train: labels from training set

batch_size: the size of one batch

epochs: the number of iterations

verbose: the level of output you want

callbacks: call a function after every epoch

validation_split: validation percentage split if you have not created it explicitly

validation_data: validation data if you have created it explicitly

Loss functions evaluate the progress of models and adjust their weights on every run. However, loss functions only describe the relationship between training data and validation data. In order to evaluate if a model is performing correctly, we typically use a third set of data—which is not used to train the network—and compare the predictions made by our model to the values available in that set of data. That is the role of the test set.

Keras provides the **model.evaluate()** method, which makes the process of evaluating a trained neural network against a test set easy. The following code illustrates how to use the **evaluate()** method:

```
model.evaluate(x=X_test, y=Y_test)
```

The **evaluate()** method returns both the results of the loss function and the results of the functions passed to the **metrics** parameter. We will be using that function frequently in the Bitcoin problem to test how the model performs on the test set.

You will notice that the Bitcoin model we trained previously looks a bit different than this example. That is because we are using an LSTM architecture. LSTMs are designed to predict sequences.

Because of that, we do not use a set of variables to predict a different single variable—even if it is a regression problem. Instead, we use previous observations from a single variable (or set of variables) to predict future observations of that same variable (or set). The **y** parameter on **keras.fit()** contains the same variable as the **x** parameter, but only the predicted sequences. So, let's have a look at how to evaluate the bitcoin model we trained previously.

EVALUATING THE BITCOIN MODEL

We created a test set during our activities in *Chapter 1, Introduction to Neural Networks and Deep Learning*. That test set contains 21 weeks of daily Bitcoin price observations, which is equivalent to about 10 percent of the original dataset.

We also trained our neural network using the other 90 percent of data (that is, the train set with 187 weeks of data, minus 1 for the validation set) in *Chapter 2, Real-World Deep Learning: Predicting the Price of Bitcoin*, and stored the trained network on disk (**bitcoin_lstm_v0**). We can now use the **evaluate()** method in each of the 21 weeks of data from the test set and inspect how that first neural network performs.

In order to do that, though, we have to provide 186 preceding weeks. We have to do this because our network has been trained to predict one week of data using exactly 186 weeks of continuous data (we will deal with this behavior by retraining our network periodically with larger periods in *Chapter 4, Productization*, when we deploy a neural network as a web application). The following snippet implements the **evaluate()** method to evaluate the performance of our model in a test dataset:

```
combined_set = np.concatenate((train_data, test_data), axis=1) \
            evaluated_weeks = []
for i in range(0, validation_data.shape[1]):
    input_series = combined_set[0:,i:i+187]

X_test = input_series[0:,:-1].reshape(1, \
            input_series.shape[1] - 1, ) \
            Y_test = input_series[0:,-1:][0]

result = B.model.evaluate(x=X_test, y=Y_test, verbose=0) \
            evaluated_weeks.append(result)
```

In the preceding code, we evaluate each week using Keras' **model.evaluate()** method, then store its output in the **evaluated_weeks** variable. We then plot the resulting MSE for each week, as shown in the following plot:

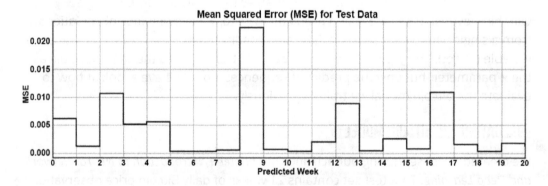

Figure 3.6: MSE for each week in the test set

The resulting MSE from our model suggests that our model performs well during most weeks, except for weeks 2, 8, 12, and 16, when its value increases from about 0.005 to 0.02. Our model seems to be performing well for almost all of the other test weeks.

OVERFITTING

Our first trained network (**bitcoin_lstm_v0**) may be suffering from a phenomenon known as **overfitting**. Overfitting is when a model is trained to optimize a validation set, but it does so at the expense of more generalizable patterns from the phenomenon we are interested in predicting. The main issue with overfitting is that a model learns how to predict the validation set, but fails to predict new data.

The loss function we used in our model reaches very low levels at the end of our training process. Not only that, but this happens early: the MSE loss function that's used to predict the last week in our data decreases to a stable plateau around epoch 30. This means that our model is predicting the data from week 187 almost perfectly, using the preceding 186 weeks. Could this be the result of overfitting?

Let's look at the preceding plot again. We know that our LSTM model reaches extremely low values in our validation set (about 2.9×10^{-6}), yet it also reaches low values in our test set. The key difference, however, is in the scale. The MSE for each week in our test set is about 4,000 times bigger (on average) than in the test set. This means that the model is performing much worse in our test data than in the validation set. This is worth considering.

The scale, though, hides the power of our LSTM model; even performing much worse in our test set, the predictions' MSE errors are still very, very low. This suggests that our model may be learning patterns from the data.

MODEL PREDICTIONS

It's one thing is to measure our model comparing MSE errors, and another to be able to interpret its results intuitively.

Using the same model, let's create a series of predictions for the following weeks, using 186 weeks as input. We do that by sliding a window of 186 weeks over the complete series (that is, train plus test sets) and making predictions for each of those windows.

The following snippet makes predictions for all the weeks of the test dataset using the **Keras model.predict()** method:

```
combined_set = np.concatenate((train_data, test_data), \
                axis=1) predicted_weeks = []
for i in range(0, validation_data.shape[1] + 1):
    input_series = combined_set[0:,i:i+186]
    predicted_weeks.append(B.predict(input_series))
```

In the preceding code, we make predictions using the **model.predict()** method, then store these predictions in the **predicted_weeks** variable. Then, we plot the resulting predictions, making the following plot:

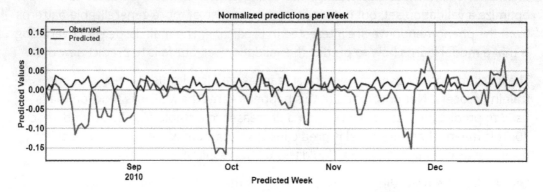

Figure 3.7: MSE for each week in the test set

The results of our model suggest that its performance isn't all that bad. By observing the pattern from the **Predicted** line (grey), we can see that the network has identified a fluctuating pattern happening on a weekly basis, in which the normalized prices go up in the middle of the week, then down by the end of it but. However, there's still a lot of room for improvement as it is unable to pick up higher fluctuations. With the exception of a few weeks—the same as with our previous MSE analysis—most weeks fall close to the correct values.

Now, let's denormalize the predictions so that we can investigate the prediction values using the same scale as the original data (that is, US dollars). We can do this by implementing a denormalization function that uses the day index from the predicted data to identify the equivalent week on the test data. After that week has been identified, the function then takes the first value of that week and uses that value to denormalize the predicted values by using the same point-relative normalization technique but inverted. The following snippet denormalizes data using an inverted point-relative normalization technique:

```
def denormalize(reference, series, normalized_variable=\
                'close_point_relative_normalization', \
                denormalized_variable='close'):

    if('iso_week' in list(series.columns)):

        week_values = reference[reference['iso_week'] \
                == series['iso_week'].values[0]]
```

```
      last_value = week_values[denormalized_variable].values[0]
      series[denormalized_variable] = \
      last_value * (series[normalized_variable] + 1)


   return series
predicted_close = predicted.groupby('iso_week').apply(lambda x: \
             denormalize(observed, x))
```

The **denormalize()** function takes the first closing price from the test's first day of an equivalent week.

Our results now compare the predicted values with the test set using US dollars. As shown in the following plot, the **bitcoin_lstm_v0** model seems to perform quite well in predicting the Bitcoin prices for the following 7 days. But how can we measure that performance in interpretable terms?

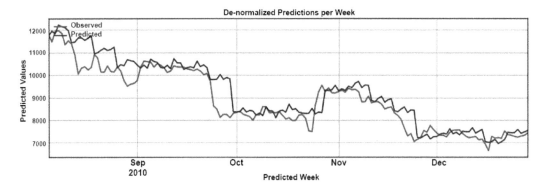

Figure 3.8: De-normalized predictions per week

INTERPRETING PREDICTIONS

Our last step is to add interpretability to our predictions. The preceding plot seems to show that our model prediction matches the test data somewhat closely, but how closely?

Keras' **model.evaluate()** function is useful for understanding how a model is performing at each evaluation step. However, given that we are typically using normalized datasets to train neural networks, the metrics that are generated by the **model.evaluate()** method are also hard to interpret.

In order to solve that problem, we can collect the complete set of predictions from our model and compare it with the test set using two other functions from *Figure 3.1* that are easier to interpret: MAPE and RMSE, which are implemented as **mape()** and **rmse()**, respectively.

> **NOTE**
>
> These functions are implemented using NumPy. The original implementations come from https://stats.stackexchange.com/questions/58391/mean-absolute-percentage-error-mape-in-scikit-learn and https://stackoverflow.com/questions/16774849/mean-squared-error-in-numpy

We can see the implementation of these methods in the following snippet:

```
def mape(A, B):
    return np.mean(np.abs((A - B) / A)) * 100

def rmse(A, B):
    return np.sqrt(np.square(np.subtract(A, B)).mean())
```

After comparing our test set with our predictions using both of those functions, we have the following results:

- **Denormalized RMSE**: $596.6 USD

- **Denormalized MAPE**: 4.7 percent

This indicates that our predictions differ, on average, about $596 from real data. This represents a difference of about 4.7 percent from real Bitcoin prices.

These results facilitate the understanding of our predictions. We will continue to use the **model.evaluate()** method to keep track of how our LSTM model is improving, but will also compute both **rmse()** and **mape()** on the complete series on every version of our model to interpret how close we are to predicting Bitcoin prices.

EXERCISE 3.01: CREATING AN ACTIVE TRAINING ENVIRONMENT

In this exercise, we'll create a training environment for our neural network that facilitates both its training and evaluation. This environment is particularly important for the next topic, in which we'll search for an optimal combination of hyperparameters.

First, we will start both a Jupyter Notebook instance and a TensorBoard instance. Both of these instances can remain open for the remainder of this exercise. Let's get started:

1. Using your Terminal, navigate to the **Chapter03/Exercise3.01** directory and execute the following code to start a Jupyter Notebook instance:

```
$ jupyter-lab
```

The server should open in your browser automatically.

2. Open the Jupyter Notebook named **Exercise3.01_Creating_an_active_training_environment.ipynb**:

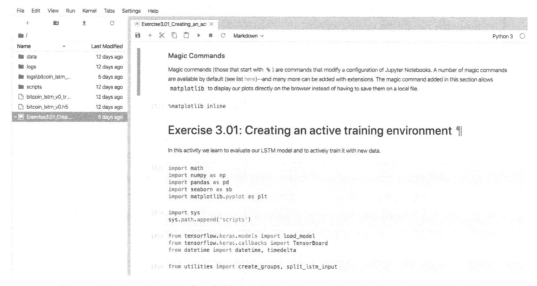

Figure 3.9: Jupyter Notebook highlighting the section, Evaluate LSTM Model

3. Also using your Terminal, start a TensorBoard instance by executing the following command:

```
$ cd ./Chapter03/Exercise3.01/
$ tensorboard --logdir=logs/
```

Ensure the **logs** directory is empty in the repository.

4. Open the URL that appears on screen and leave that browser tab open, as well. Execute the initial cells containing the import statements to ensure that the dependencies are loaded.

5. Execute the two cells under Validation Data to load the train and test datasets in the Jupyter Notebook:

```
train = pd.read_csv('data/train_dataset.csv')
test = pd.read_csv('data/test_dataset.csv')
```

> **NOTE**
>
> Don't forget to change the path (highlighted) of the files based on where they are saved on your system.

6. Add TensorBoard callback and retrain the model. Execute the cells under Re-Train model with TensorBoard.

Now, let's evaluate how our model performed against the test data. Our model is trained using 186 weeks to predict a week into the future—that is, the following sequence of 7 days. When we built our first model, we divided our original dataset between a training and a test set. Now, we will take a combined version of both datasets (let's call it a combined set) and move a sliding window of 186 weeks. At each window, we execute Keras' **model.evaluate()** method to evaluate how the network performed on that specific week.

7. Execute the cells under the header, **Evaluate LSTM Model**. The key concept of these cells is to call the **model.evaluate()** method for each of the weeks in the test set. This line is the most important:

```
result = model.evaluate(x=X_test, y=Y_test, verbose=0)
```

8. Each evaluation result is now stored in the **evaluated_weeks** variable. That variable is a simple array containing the sequence of MSE predictions for every week in the test set. Go ahead and plot the results:

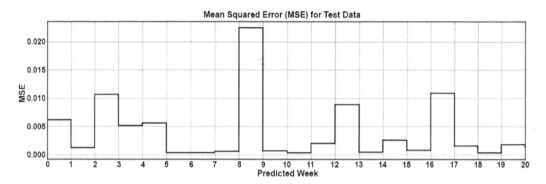

Figure 3.10: MSE results from the model.evaluate() method for each week of the test set

As we've already discussed, the MSE loss function is difficult to interpret. To facilitate our understanding of how our model is performing, we also call the **model.predict()** method on each week from the test set and compare its predicted results with the set's values.

9. Navigate to the *Interpreting Model Results* section and execute the code cells under the **Make Predictions** subheading. Notice that we are calling the **model.predict()** method, but with a slightly different combination of parameters. Instead of using both the **X** and **Y** values, we only use **X**:

```
predicted_weeks = []
for i in range(0, test_data.shape[1]):
    input_series = combined_set[0:,i:i+186]
    predicted_weeks.append(model.predict(input_series))
```

At each window, we will issue predictions for the following week and store the results. Now, we can plot the normalized results alongside the normalized values from the test set, as shown in the following plot:

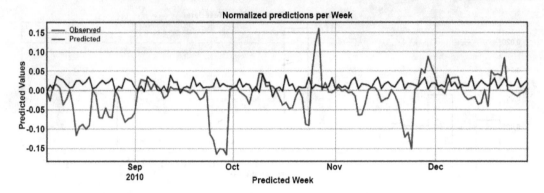

Figure 3.11: Plotting the normalized values returned from model.predict() for each week of the test set

We will also make the same comparisons but using denormalized values. In order to denormalize our data, we must identify the equivalent week between the test set and the predictions. Then, we can take the first price value for that week and use it to reverse the point-relative normalization equation from *Chapter 2, Real-World Deep Learning: Predicting the Price of Bitcoin*.

10. Navigate to the **De-normalized Predictions** header and execute all the cells under that header.

11. In this section, we defined the **denormalize()** function, which performs the complete denormalization process. In contrast to the other functions, this function takes in a pandas DataFrame instead of a NumPy array. We do this to use dates as an index. This is the most relevant cell block from that header:

```
predicted_close = predicted.groupby('iso_week').apply(\
                    lambda x: denormalize(observed, x))
```

Our denormalized results (as seen in the following plot) show that our model makes predictions that are close to the real Bitcoin prices. But how close?

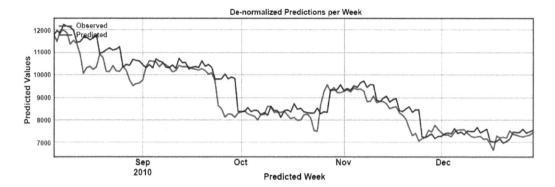

Figure 3.12: Plotting the denormalized values returned from model.predict()
for each week of the test set

The LSTM network uses MSE values as its loss function. However, as we've already discussed, MSE values are difficult to interpret. To solve this, we need to implement two functions (loaded from the **utilities.py** script) that implement the **rmse()** and **mape()** functions. These functions add interpretability to our model by returning a measurement on the same scale that our original data used, and by comparing the difference in scale as a percentage.

12. Navigate to the **De-normalizing Predictions** header and load two functions from the **utilities.py** script:

```
from scripts.utilities import rmse, mape
```

The functions from this script are actually really simple:

```
def mape(A, B):
    return np.mean(np.abs((A - B) / A)) * 100
def rmse(A, B):
    return np.sqrt(np.square(np.subtract(A, B)).mean())
```

Each function is implemented using NumPy's vector-wise operations. They work well in vectors of the same length. They are designed to be applied on a complete set of results.

Using the **mape()** function, we can now understand that our model predictions are about 4.7 percent away from the prices from the test set. This is equivalent to a RSME (calculated using the **rmse()** function) of about $596.6.

Before moving on to the next section, go back into the Notebook and find the **Re-train Model with TensorBoard** header. You may have noticed that we created a helper function called **train_model()**. This function is a wrapper around our model that trains (using **model.fit()**) our model, storing its respective results under a new directory. Those results are then used by TensorBoard in order to display statistics for different models.

13. Go ahead and modify some of the values for the parameters that were passed to the **model.fit()** function (try epochs, for instance). Now, run the cells that load the model into memory from disk (this will replace your trained model):

```
model = load_model('bitcoin_lstm_v0.h5')
```

14. Now, run the **train_model()** function again, but with different parameters, indicating a new run version. When you run this command, you will be able to train a newer version of the model and specify the newer version in the version parameter:

```
train_model(model=model, X=X_train, Y=Y_train, \
            epochs=10, version=0, run_number=0)
```

> **NOTE**
>
> To access the source code for this specific section, please refer to https://packt.live/2ZhK4z3.
>
> You can also run this example online at https://packt.live/2Dvd9i3.
> You must execute the entire Notebook in order to get the desired result.

In this exercise, we learned how to evaluate a network using loss functions. We learned that loss functions are key elements of neural networks since they evaluate the performance of a network at each epoch and are the starting point for the propagation of adjustments back into layers and nodes. We also explored why some loss functions can be difficult to interpret (for instance, the MSE function) and developed a strategy using two other functions—RMSE and MAPE—to interpret the predicted results from our LSTM model.

Most importantly, we've concluded this exercise with an active training environment. We now have a system that can train a deep learning model and evaluate its results continuously. This will be key when we move on to optimizing our network in the next topic.

HYPERPARAMETER OPTIMIZATION

So far, we have trained a neural network to predict the next 7 days of Bitcoin prices using the preceding 76 weeks of prices. On average, this model issues predictions that are about 8.4 percent distant from real Bitcoin prices.

This section describes common strategies for improving the performance of neural network models:

- Adding or removing layers and changing the number of nodes

- Increasing or decreasing the number of training epochs

- Experimenting with different activation functions

- Using different regularization strategies

We will evaluate each modification using the same active learning environment we developed by the end of the *Model Evaluation* section, measuring how each one of these strategies may help us develop a more precise model.

LAYERS AND NODES – ADDING MORE LAYERS

Neural networks with single hidden layers can perform fairly well on many problems. Our first Bitcoin model (**bitcoin_lstm_v0**) is a good example: it can predict the next 7 days of Bitcoin prices (from the test set) with error rates of about 8.4 percent using a single LSTM layer. However, not all problems can be modeled with single layers.

The more complex the function you are working to predict, the higher the likelihood that you will need to add more layers. A good way to determine whether adding new layers is a good idea is to understand what their role in a neural network is.

Each layer creates a model representation of its input data. Earlier layers in the chain create lower-level representations, while later layers create higher-level representations.

While this description may be difficult to translate into real-world problems, its practical intuition is simple: when working with complex functions that have different levels of representation, you may want to experiment with adding layers.

ADDING MORE NODES

The number of neurons that your layer requires is related to how both the input and output data is structured. For instance, if you are working on a binary classification problem to classify a 4 x 4 pixel image, then you can try out the following:

- Have a hidden layer that has 12 neurons (one for each available pixel)

- Have an output layer that has only two neurons (one for each predicted class)

It is common to add new neurons alongside the addition of new layers. Then, we can add a layer that has either the same number of neurons as the previous one, or a multiple of the number of neurons from the previous layer. For instance, if your first hidden layer has 12 neurons, you can experiment with adding a second layer that has either 12, 6, or 24 neurons.

Adding layers and neurons can have significant performance limitations. Feel free to experiment with adding layers and nodes. It is common to start with a smaller network (that is, a network with a small number of layers and neurons), then grow according to its performance gains.

If this comes across as imprecise, your intuition is right.

> **NOTE**
>
> To quote *Aurélien Géron*, YouTube's former lead for video classification, "*Finding the perfect amount of neurons is still somewhat of a black art.*"

Finally, a word of caution: the more layers you add, the more hyperparameters you have to tune—and the longer your network will take to train. If your model is performing fairly well and not overfitting your data, experiment with the other strategies outlined in this chapter before adding new layers to your network.

LAYERS AND NODES – IMPLEMENTATION

Now, we will modify our original LSTM model by adding more layers. In LSTM models, we typically add LSTM layers in a sequence, making a chain between LSTM layers. In our case, the new LSTM layer has the same number of neurons as the original layer, so we don't have to configure that parameter.

We will name the modified version of our model **bitcoin_lstm_v1**. It is good practice to name each one of the models in terms of which one is attempting different hyperparameter configurations. This helps you to keep track of how each different architecture performs, and also to easily compare model differences in TensorBoard. We will compare all the different modified architectures at the end of this chapter.

> **NOTE**
>
> Before adding a new LSTM layer, we need to set the **return_sequences** parameter to **True** on the first LSTM layer. We do this because the first layer expects a sequence of data with the same input as that of the first layer. When this parameter is set to **False**, the LSTM layer outputs the predicted parameters in a different, incompatible output.

The following code example adds a second LSTM layer to the original **bitcoin_lstm_v0** model, making it **bitcoin_lstm_v1**:

```
period_length = 7
number_of_periods = 186
batch_size = 1

model = Sequential()
model.add(LSTM(
units=period_length,
batch_input_shape=(batch_size, number_of_periods, period_length), \
                input_shape=(number_of_periods, period_length), \
                return_sequences=True, stateful=False))

model.add(LSTM(units=period_length, \
              batch_input_shape=(batch_size, number_of_periods, \
                                 period_length), \
              input_shape=(number_of_periods, period_length), \
              return_sequences=False, stateful=False))

model.add(Dense(units=period_length)) \
model.add(Activation("linear"))

model.compile(loss="mse", optimizer="rmsprop")
```

EPOCHS

Epochs are the number of times the network adjusts its weights in response to the data passing through and its resulting loss function. Running a model for more epochs can allow it to learn more from data, but you also run the risk of overfitting.

When training a model, prefer to increase the epochs exponentially until the loss function starts to plateau. In the case of the **bitcoin_lstm_v0** model, its loss function plateaus at about 100 epochs.

Our LSTM model uses a small amount of data to train, so increasing the number of epochs does not affect its performance in a significant way. For instance, if we attempt to train it at 103 epochs, the model barely gains any improvements. This will not be the case if the model being trained uses enormous amounts of data. In those cases, a large number of epochs is crucial to achieve good performance.

I suggest you use the following rule of thumb: *the larger the data used to train your model, the more epochs it will need to achieve good performance.*

EPOCHS – IMPLEMENTATION

Our Bitcoin dataset is rather small, so increasing the epochs that our model trains may have only a marginal effect on its performance. In order to have the model train for more epochs, we only have to change the **epochs** parameter in the **model. fit()** method. In the following snippet, you will see how to change the number of epochs that our model trains for:

```
number_of_epochs = 10**3
model.fit(x=X, y=Y, batch_size=1, \
          epochs=number_of_epochs, \
          verbose=0, \
          callbacks=[tensorboard])
```

This change bumps our model to **v2**, effectively making it **bitcoin_lstm_v2**.

ACTIVATION FUNCTIONS

Activation functions evaluate how much you need to activate individual neurons. They determine the value that each neuron will pass to the next element of the network, using both the input from the previous layer and the results from the loss function—or if a neuron should pass any values at all.

> **NOTE**
>
> Activation functions are a topic of great interest for those in the scientific community researching neural networks. For an overview of research currently being done on the topic and a more detailed review on how activation functions work, please refer to *Deep Learning by Ian Goodfellow et. al., MIT Press, 2017.*

TensorFlow and Keras provide many activation functions—and new ones are occasionally added. As an introduction, three are important to consider; let's explore each of them.

> **NOTE**
>
> This section has been greatly inspired by the article *Understanding Activation Functions in Neural Networks* by Avinash Sharma V, available at https://medium.com/the-theory-of-everything/understanding-activation-functions-in-neural-networks-9491262884e0.

LINEAR (IDENTITY) FUNCTIONS

Linear functions only activate a neuron based on a constant value. They are defined by the following equation:

$$f(x) = c * (0, x)$$

Figure 3.13: Formula for linear functions

Here, c is the constant value. When c = 1, neurons will pass the values as is, without any modification needed by the activation function. The issue with using linear functions is that, due to the fact that neurons are activated linearly, chained layers now function as a single large layer. In other words, we lose the ability to construct networks with many layers, in which the output of one influences the other:

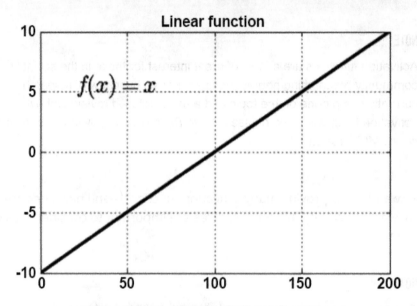

Figure 3.14: Illustration of a linear function

The use of linear functions is generally considered obsolete for most networks because they do not compute complex features and do not induce proper non-linearity in neurons.

HYPERBOLIC TANGENT (TANH) FUNCTION

Tanh is a non-linear function, and is represented by the following formula:

$$f(x) = \frac{2}{2 + e^{-2x}} - 1$$

Figure 3.15: Formula for hyperbolic tangent function

This means that the effect they have on nodes is evaluated continuously. Also, because of its non-linearity, we can use this function to change how one layer influences the next layer in the chain. When using non-linear functions, layers activate neurons in different ways, making it easier to learn different representations from data. However, they have a sigmoid-like pattern that penalizes extreme node values repeatedly, causing a problem called vanishing gradients. Vanishing gradients have negative effects on the ability of a network to learn:

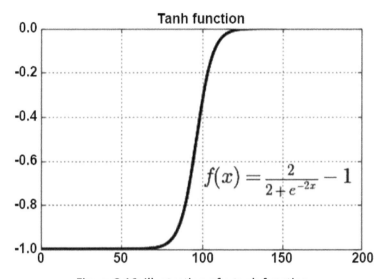

Figure 3.16: Illustration of a tanh function

Tanhs are popular choices, but due to the fact that they are computationally expensive, ReLUs are often used instead.

RECTIFIED LINEAR UNIT FUNCTIONS

ReLU stands for **Rectified Linear Unit**. It filters out negative values and keeps only the positive values. ReLU functions are often recommended as great starting points before trying other functions. They are defined by the following formula:

$$f(x) = max(0, x)$$

Figure 3.17: Formula for ReLU functions

ReLUs have non-linear properties:

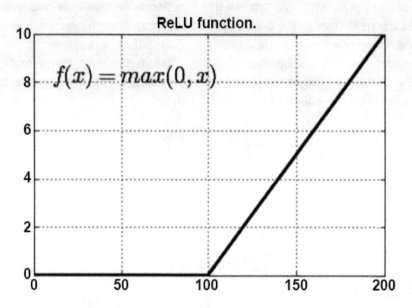

Figure 3.18: Illustration of a ReLU function

ReLUs tend to penalize negative values. So, if the input data (for instance, normalized between -1 and 1) contains negative values, those will now be penalized by ReLUs. That may not be the intended behavior.

We will not be using ReLU functions in our network because our normalization process creates many negative values, yielding a much slower learning model.

ACTIVATION FUNCTIONS – IMPLEMENTATION

The easiest way to implement activation functions in Keras is by instantiating the **Activation()** class and adding it to the **Sequential()** model. **Activation()** can be instantiated with any activation function available in Keras (for a complete list, see https://keras.io/activations/).

In our case, we will use the **tanh** function. After implementing an activation function, we bump the version of our model to **v2**, making it **bitcoin_lstm_v3**:

```
model = Sequential() model.add(LSTM(
                        units=period_length,\
                        batch_input_shape=(batch_size, \
                        number_of_periods, period_length), \
                        input_shape=(number_of_periods, \
                                period_length), \
```

```
                          return_sequences=True, \
                          stateful=False))

model.add(LSTM(units=period_length,\
          batch_input_shape=(batch_size, number_of_periods, \
                             period_length), \
          input_shape=(number_of_periods, period_length), \
          return_sequences=False, stateful=False))

model.add(Dense(units=period_length)) \
model.add(Activation("tanh"))

model.compile(loss="mse", optimizer="rmsprop")
```

After executing the **compile** command, your model has been built according to the layers specified and is now ready to be trained. There are a number of other activation functions worth experimenting with. Both TensorFlow and Keras provide a list of implemented functions in their respective official documentations. Before implementing your own, start with the ones we've already implemented in both TensorFlow and Keras.

REGULARIZATION STRATEGIES

Neural networks are particularly prone to overfitting. Overfitting happens when a network learns the patterns of the training data but is unable to find generalizable patterns that can also be applied to the test data.

Regularization strategies refer to techniques that deal with the problem of overfitting by adjusting how the network learns. In the following sections, we'll discuss two common strategies:

- L2 Regularization
- Dropout

L2 REGULARIZATION

L2 regularization (or **weight decay**) is a common technique for dealing with overfitting models. In some models, certain parameters vary in great magnitudes. L2 regularization penalizes such parameters, reducing the effect of these parameters on the network.

L2 regularizations use the λ parameter to determine how much to penalize a model neuron. We typically set that to a very low value (that is, 0.0001); otherwise, we risk eliminating the input from a given neuron completely.

DROPOUT

Dropout is a regularization technique based on a simple question: *if we randomly take away a proportion of the nodes from the layers, how will the other node adapt?* It turns out that the remaining neurons adapt, learning to represent patterns that were previously handled by those neurons that are missing.

The dropout strategy is simple to implement and is typically very effective at avoiding overfitting. This will be our preferred regularization.

REGULARIZATION STRATEGIES – IMPLEMENTATION

In order to implement the dropout strategy using Keras, we'll import the **Dropout()** method and add it to our network immediately after each LSTM layer. This addition effectively makes our network **bitcoin_lstm_v4**. In this snippet, we're adding the **Dropout()** step to our model (**bitcoin_lstm_v3**), making it **bitcoin_lstm_v4**:

```
model = Sequential()
model.add(LSTM(\
        units=period_length,\
        batch_input_shape=(batch_size, number_of_periods, \
                        period_length), \
        input_shape=(number_of_periods, period_length), \
        return_sequences=True, stateful=False))
model.add(Dropout(0.2))
model.add(LSTM(\
        units=period_length,\
        batch_input_shape=(batch_size, number_of_periods, \
                        period_length), \
        input_shape=(number_of_periods, period_length), \
        return_sequences=False, stateful=False))

model.add(Dropout(0.2))

model.add(Dense(units=period_length))
```

```
model.add(Activation("tanh"))

model.compile(loss="mse", optimizer="rmsprop")
```

We could have used L2 regularization instead of dropout. Dropout drops out random neurons in each epoch, whereas L2 regularization penalizes neurons that have high weight values. In order to apply L2 regularization, simply instantiate the **ActivityRegularization()** class with the L2 parameter set to a low value (for instance, 0.0001). Then, place it in the place where the **Dropout()** class has been added to the network. Feel free to experiment by adding that to the network while keeping both **Dropout()** steps, or simply replace all the **Dropout()** instances with **ActivityRegularization()** instead.

OPTIMIZATION RESULTS

All in all, we have created four versions of our model. Three of these versions, that is, **bitcoin_lstm_v1**, **bitcoin_lstm_v2**, and **bitcoin_lstm_v3**, were created by applying different optimization techniques that were outlined in this chapter. Now, we have to evaluate which model performs best. In order to do that, we will use the same metrics we used in our first model: MSE, RMSE, and MAPE. MSE is used to compare the error rates of the model on each predicted week. RMSE and MAPE are computed to make the model results easier to interpret. The following table shows this:

Model	RMSE (whole series)	MAPE (whole series)	Training Time
bitcoin_lstm_v0	667	5.3 percent	-
bitcoin_lstm_v1	1196	10.3 percent	10.3 s
bitcoin_lstm_v2	1185	10 percent	1 min 8s
bitcoin_lstm_v3	1050	8.1 percent	27.2s
bitcoin_lstm_v4	1195	10 percent	2 min 11s

Figure 3.19: Model results for all models

Interestingly, our first model (**bitcoin_lstm_v0**) performed the best in nearly all defined metrics. We will be using that model to build our web application and continuously predict Bitcoin prices.

ACTIVITY 3.01: OPTIMIZING A DEEP LEARNING MODEL

In this activity, we'll implement different optimization strategies on the model we created in *Chapter 2, Real-World Deep Learning: Predicting the Price of Bitcoin* (`bitcoin_lstm_v0`). This model achieves a MAPE performance on the complete de-normalization test set of about 8.4 percent. We will try to reduce that gap and get more accurate predictions.

Here are the steps:

1. Start TensorBoard from a Terminal.

2. Start a Jupyter Notebook.

3. Load the train and test data and split the `lstm` input in the format required by the model.

4. In the previous exercise, we create a model architecture. Copy that model architecture and add a new LSTM layer. Compile and create a model.

5. Change the number of epochs in *step 4* by creating a new model. Compile and create a new model.

6. Change the activation function to **tanh** or **relu** and create a new model. Compile and train a new model.

7. Add a new layer for dropout after the LSTM layer and create a new model. Keep values such as **0.2** or **0.3** for dropout. Compile and train a new model.

8. Evaluate and compare all the models that were trained in this activity.

> ### NOTE
> The solution to this activity can be found on page 141.

SUMMARY

In this chapter, we learned how to evaluate our model using the MSE, RMSE, and MAPE metrics. We computed the latter two metrics in a series of 19-week predictions made by our first neural network model. By doing this, we learned that it was performing well.

We also learned how to optimize a model. We looked at optimization techniques, which are typically used to increase the performance of neural networks. Also, we implemented a number of these techniques and created a few more models to predict Bitcoin prices with different error rates.

In the next chapter, we will be turning our model into a web application that does two things: retrains our model periodically with new data and is able to make predictions using an HTTP API interface.

4

PRODUCTIZATION

OVERVIEW

In this chapter, you will handle new data and create a model that is able to learn continuously from the patterns it is shown and help make better predictions. We will use a web application as an example to show how to deploy deep learning models not only because of the simplicity and prevalence of web apps, but also the different possibilities they provide, such as getting predictions on mobile using a web browser and making REST APIs for users.

INTRODUCTION

This chapter focuses on how to *productize* a deep learning model. We use the word productize to define the creation of a software product from a deep learning model that can be used by other people and applications.

We are interested in models that use new data as and when it becomes available, continuously learn patterns from new data, and consequently, make better predictions. In this chapter, we will study two strategies to deal with new data: one that retrains an existing model, and another that creates a completely new model. Then, we implement the latter strategy in our Bitcoin price prediction model so that it can continuously predict new Bitcoin prices.

By the end of this chapter, we will be able to deploy a working web application (with a functioning HTTP API) and modify it to our heart's content.

HANDLING NEW DATA

Models can be trained once using a set of data and can then be used to make predictions. Such static models can be very useful, but it is often the case that we want our model to continuously learn from new data—and to continuously get better as it does so.

In this section, we will discuss two strategies of handling new data and see how to implement them in Python.

SEPARATING DATA AND MODEL

When building a deep learning application, the two most important areas are data and model. From an architectural point of view, it is recommended that these two areas be kept separate. We believe that is a good suggestion because each of these areas includes functions inherently separate from each other. Data is often required to be collected, cleaned, organized, and normalized, whereas models need to be trained, evaluated, and able to make predictions.

Following that suggestion, we will be using two different code bases to help us build our web application: the Yahoo Finance API and **Model()**:

- The Yahoo Finance API: The API can be installed by using **pip** with the following command:

```
pip install yfinance
```

After installation, we will be able to access all the historical data related to the finance domain.

- **Model()**: This class implements all the code we have written so far into a single class. It provides facilities for interacting with our previously trained models and allows us to make predictions using de-normalized data, which is much easier to understand. The **Model()** class is our model component.

These two code bases are used extensively throughout our example application and define the data and model components.

THE DATA COMPONENT

The Yahoo Finance API helps to retrieve and parse the historical data of stocks. It contains one relevant method, **history()**, which is detailed in the following code:

```
import yfinance as yf
ticker =  yf.Ticker("BTC-USD")
historic_data = ticker.history(period='max')
```

This **history()** method collects data from the Yahoo Finance website, parses it, and returns a pandas DataFrame that is ready to be used by the **Model()** class.

The Yahoo Finance API uses the parameter ticker to determine what cryptocurrency to collect. The Yahoo Finance API has many other cryptocurrencies available, including popular ones such as Ethereum and Bitcoin Cash. Using the **ticker** parameter, you can change the cryptocurrency and train a different model apart from the Bitcoin model created in this book.

THE MODEL COMPONENT

The **Model()** class is where we implement the application's model component. The **Model()** class contains five methods that implement all the different modeling topics from this book. They are the following:

- **build()**: This method builds an LSTM model using TensorFlow. This method works as a simple wrapper for a manually created model.

- **train()**: This method trains the model using data that the class was instantiated with.

- **evaluate()**: This method evaluates the model using a set of loss functions.

- **save()**: This method saves the model locally as a file.

- **predict()**: This method makes and returns predictions based on an input sequence of observations ordered by week.

We will use these methods throughout this chapter to work, train, evaluate, and issue predictions with our model. The **Model()** class is an example of how to wrap essential TensorFlow functions into a web application. The preceding methods can be implemented almost exactly as in previous chapters, but with enhanced interfaces. For example, the **train()** method implemented in the following code trains a model available in **self.model** using data from **self.X** and **self.Y**:

```
def train(self, data=None, epochs=300, verbose=0, batch_size=1):
    self.train_history = self.model.fit(x=self.X, y=self.Y, \
                                        batch_size=batch_size, \
                                        epochs=epochs, \
                                        verbose=verbose, \
                                        shuffle=False)

    self.last_trained = datetime.now()\
    .strftime('%Y-%m-%d %H:%M:%S')
    return self.train_history
```

The general idea is that each of the processes from the Keras workflow (build or design, train, evaluate, and predict) can easily be turned into distinct parts of a program. In our case, we have made them into methods that can be invoked from the **Model()** class. This organizes our program and provides a series of constraints (such as on the model architecture or certain API parameters), which help us deploy our model in a stable environment.

In the following sections, we will explore common strategies for dealing with new data.

DEALING WITH NEW DATA

The core idea of machine learning models—neural networks included—is that they can learn patterns from data. Imagine that a model was trained with a certain dataset and it is now issuing predictions. Now, imagine that new data is available. There are different strategies you can employ so that a model can take advantage of the newly available data to learn new patterns and improve its predictions. In this section, we will discuss two strategies:

- Retraining an old model

- Training a new model

RETRAINING AN OLD MODEL

In this strategy, we retrain an existing model with new data. Using this strategy, you can continuously adjust the model parameters to adapt to new phenomena. However, data used in later training periods might be significantly different from earlier data. Such differences might cause significant changes to the model parameters, such as making it learn new patterns and forget old patterns. This phenomenon is generally referred to as **catastrophic forgetting**.

> **NOTE**
>
> Catastrophic forgetting is a common phenomenon affecting neural networks. Deep learning researchers have been trying to tackle this problem for many years. DeepMind, a Google-owned deep learning research group from the United Kingdom, has made notable advancements in finding a solution. The article, *Overcoming Catastrophic Forgetting in Neural Networks, James Kirkpatrick, et. al.* is a good reference for such work, and is available at https://arxiv.org/pdf/1612.00796.pdf.

The interface used for training (**model.fit()**) for the first time can be used for training with new data as well. The following snippet loads the data and helps to train a model specifying the epochs and batch size:

```
X_train_new, Y_train_new = load_new_data()

model.fit(x=X_train_new, y=Y_train_new, batch_size=1, \
          epochs=100, verbose=0)
```

In TensorFlow, when models are trained, the model's state is saved as weights on the disk. When you use the **model.save()** method, that state is also saved. And when you invoke the **model.fit()** method, the model is retrained with the new dataset, using the previous state as a starting point.

In typical Keras models, this technique can be used without further issues. However, when working with LSTM models, this technique has one key limitation: the shape of both train and validation data must be the same. For example, in *Chapter 3, Real-World Deep Learning: Evaluating the Bitcoin Model*, our LSTM model (**bitcoin_lstm_v0**) uses 186 weeks to predict one week into the future. If we attempt to retrain the model with 187 weeks to predict the coming week, the model raises an exception with information regarding the incorrect shape of data.

One way of dealing with this is to arrange data in the format expected by the model. For example, to make predictions based on a year's data (52 weeks), we would need to configure a model to predict a future week using 40 weeks. In this case, we first train the model with the first 40 weeks of 2019, then continue to retrain it over the following weeks until we reach week 51. We use the **Model()** class to implement a retraining technique in the following code:

```
M = Model(data=model_data[0*7:7*40 + 7], variable='close', \
          predicted_period_size=7)


M.build()
M.train()


for i in range(41, 52):
    j = i - 40
    M.train(model_data.loc[j*7:7*i + 7])
```

This technique tends to be fast to train and tends to work well with series that are large. The next technique is easier to implement and works well in smaller series.

TRAINING A NEW MODEL

Another strategy is to create and train a new model every time new data is available. This approach tends to reduce catastrophic forgetting, but training time increases as data increases. Its implementation is quite simple.

Using the Bitcoin model as an example, let's now assume that we have old data for 49 weeks of 2019, and that after a week, new data is available. We represent this with the **old_data** and **new_data** variables in the following snippet, in which we implement a strategy for training a new model when new data is available:

```
old_data = model_data[0*7:7*48 + 7]
new_data = model_data[0*7:7*49 + 7]


M = Model(data=old_data, \
          variable='close', predicted_period_size=7)


M.build()
M.train()


M = Model(data=new_data, \
          variable='close', predicted_period_size=7)
```

```
M.build()
M.train()
```

This approach is very simple to implement and tends to work well for small datasets. This will be the preferred solution for our Bitcoin price-predictions application.

EXERCISE 4.01: RETRAINING A MODEL DYNAMICALLY

In this exercise, you have to retrain a model to make it dynamic. Whenever new data is loaded, it should be able to make predictions accordingly. Here are the steps to follow:

Start by importing **cryptonic**. Cryptonic is a simple software application developed for this book that implements all the steps up to this section using Python classes and modules. Consider Cryptonic as a template to be used to create applications. Cryptonic, provided as a Python module for this exercise, can be found at the following GitHub link: https://packt.live/325WdZQ.

1. First, we will start a Jupyter Notebook instance, and then we will load the **cryptonic** package.

2. Using your Terminal, navigate to the **Chapter04/Exercise4.01** directory and execute the following code to start a Jupyter Notebook instance:

   ```
   $ jupyter-lab
   ```

   The server will automatically open in your browser, then open the Jupyter Notebook named **Exercise4.01_Re_training_a_model_dynamically.ipynb**.

3. Now, we will import classes from the **cryptonic** package: **Model()** and the Yahoo Finance API. These classes facilitate the process of manipulating our model.

4. In the Jupyter Notebook instance, navigate to the header **Fetching Real-Time Data**. We will now be fetching updated historical data from Yahoo Finance by calling the **history()** method:

   ```
   import yfinance as yf
   ticker =  yf.Ticker("BTC-USD")
   historic_data = ticker.history(period='max')
   ```

The **historic_data** variable is now populated with a pandas DataFrame that contains historic data of Bitcoin rates up to the time of running this code. This is great and makes it easier to retrain our model when more data is available.

5. You can view the first three rows of data stored in **historic_data** using the following command:

```
historic_data.head(3)
```

You can then view this data stored in **historic_data**:

```
[9]: historic_data.head(3)
```

[9]:		date	open	high	low	close	volume
	0	2014-09-17	465.86	468.17	452.42	457.33	21056800
	1	2014-09-18	456.86	456.86	413.10	424.44	34483200
	2	2014-09-19	424.10	427.83	384.53	394.80	37919700

Figure 4.1: Output displaying the head of the data

The data contains practically the same variables from the Bitcoin dataset we used. However, much of the data comes from an earlier period, 2017 to 2019.

6. Using the pandas API, filter the data for only the dates available in 2019, and store them in **model_data**. You should be able to do this by using the date variable as the filtering index. Make sure the data is filtered before you continue:

```
start_date = '01-01-2019'
end_date = '31-12-2019'
mask = ((historic_data['date'] \
        >= start_date) & (historic_data['date'] \
        <= end_date))
model_data = historic_data[mask]
```

Run **model_data** in next cell and the output model can be seen as follows:

```
[11]:  model_data
```

```
[11]:
```

	date	open	high	low	close	volume
0	2019-01-01	3746.71	3850.91	3707.23	3843.52	4324200990
1	2019-01-02	3849.22	3947.98	3817.41	3943.41	5244856835
2	2019-01-03	3931.05	3935.69	3826.22	3836.74	4530215218
3	2019-01-04	3832.04	3865.93	3783.85	3857.72	4847965467
4	2019-01-05	3851.97	3904.90	3836.90	3845.19	5137609823
...	...	...	...	...	...	...
360	2019-12-27	7238.14	7363.53	7189.93	7290.09	22777360995
361	2019-12-28	7289.03	7399.04	7286.91	7317.99	21365673026
362	2019-12-29	7317.65	7513.95	7279.87	7422.65	22445257701
363	2019-12-30	7420.27	7454.82	7276.31	7293.00	22874131671
364	2019-12-31	7294.44	7335.29	7169.78	7193.60	21167946112

365 rows × 6 columns

Figure 4.2: The model_data variable showing historical data

The **Model ()** class compiles all the code we have written so far in all of our activities. We will use that class to build, train, and evaluate our model in this activity.

7. We will now use the filtered data to train the model:

```
M = Model(data=model_data, \
          variable='close', predicted_period_size=7)

M.build()
M.train()
M.predict(denormalized=True)
```

8. Run the following command to see the trained model:

```
M.train(epochs=100, verbose=1)
```

The trained model is shown in the following screenshot:

```
[18]: M.train(epochs=100, verbose=1)

Train on 1 samples
Epoch 1/100
1/1 [==============================] - 0s 397ms/sample - loss: 1.4343e-04
Epoch 2/100
1/1 [==============================] - 0s 18ms/sample - loss: 5.9645e-05
Epoch 3/100
1/1 [==============================] - 0s 20ms/sample - loss: 3.1273e-05
Epoch 4/100
1/1 [==============================] - 0s 21ms/sample - loss: 1.8176e-05
Epoch 5/100
1/1 [==============================] - 0s 34ms/sample - loss: 9.5220e-06
Epoch 6/100
1/1 [==============================] - 0s 21ms/sample - loss: 5.4068e-06
Epoch 7/100
1/1 [==============================] - 0s 21ms/sample - loss: 2.9641e-06
Epoch 8/100
1/1 [==============================] - 0s 20ms/sample - loss: 1.6260e-06
Epoch 9/100
1/1 [==============================] - 0s 23ms/sample - loss: 9.3761e-07
Epoch 10/100
1/1 [==============================] - 0s 21ms/sample - loss: 6.2995e-07
```

Figure 4.3: The output showing our trained model

The preceding steps showcase the complete workflow when using the **Model()** class to train a model.

> **NOTE**
>
> For the complete code, use the **Chapter04/Exercise4.01** folder.

9. Next, we'll focus on retraining our model every time more data is available. This readjusts the weights of the network to new data.

In order to do this, we have configured our model to predict a week using 40 weeks. We now want to use the remaining 11 full weeks to create overlapping periods of 40 weeks. These include one of those 11 weeks at a time, and retrain the model for every one of those periods.

10. Navigate to the **Re-Train Old Model** header in the Jupyter Notebook. Now, complete the **range** function and the **model_data** filtering parameters using an index to split the data into overlapping groups of seven days. Then, retrain our model and collect the results:

```
results = []
for i in range(A, B):
    M.train(model_data[C:D])
    results.append(M.evaluate())
```

The **A**, **B**, **C**, and **D** variables are placeholders. Use integers to create overlapping groups of seven days in which the overlap is of one day.

Replacing these placeholders with weeks, we run the loop as follows:

```
results = []
for i in range(41, 52):
    j = i-40
    print("Training model {0} for week {1}".format(j,i))
    M.train(model_data.loc[j*7:7*i+7])
    results.append(M.evaluate())
```

Here's the output showing the results of this loop:

```
Training model 1 for week 41
Training model 2 for week 42
Training model 3 for week 43
Training model 4 for week 44
Training model 5 for week 45
Training model 6 for week 46
Training model 7 for week 47
Training model 8 for week 48
Training model 9 for week 49
Training model 10 for week 50
Training model 11 for week 51
```

11. After you have retrained your model, go ahead and invoke the **M.predict(denormalized=True)** function and examine the results:

```
array([7187.145 , 7143.798 , 7113.7324, 7173.985 , 7200.346 ,
       7300.2896, 7175.3203], dtype=float32)
```

12. Next, we'll focus on creating and training a new model every time new data is available. In order to do this, we now assume that we have old data for 49 weeks of 2019, and after a week, we now have new data. We represent this with the **old_data** and **new_data** variables.

13. Navigate to the **New Data New Model** header and split the data between the **old_data** and **new_data** variables:

```
old_data = model_data[0*7:7*48 + 7]
new_data = model_data[0*7:7*49 + 7]
```

14. Then, train the model with **old_data** first:

```
M = Model(data=old_data,\
          variable='close', predicted_period_size=7)
M.build()
M.train()
```

We now have all the pieces that we need in order to train our model dynamically.

> **NOTE**
>
> To access the source code for this specific section, please refer to https://packt.live/2AQb3bE.
>
> You can also run this example online at https://packt.live/322KuLl.
> You must execute the entire Notebook in order to get the desired result.

In the next section, we will deploy our model as a web application, making its predictions available in the browser via an HTTP API.

DEPLOYING A MODEL AS A WEB APPLICATION

In this section, we will deploy our model as a web application. We will use the Cryptonic web application to deploy our model, exploring its architecture so that we can make modifications in the future. The intention is to have you use this application as a starter for more complex applications—a starter that is fully working and can be expanded as you see fit.

Aside from familiarity with Python, this topic assumes familiarity with creating web applications. Specifically, we assume that you have some knowledge of web servers, routing, the HTTP protocol, and caching. You will be able to locally deploy the demonstrated Cryptonic application without extensive knowledge of these web servers, the HTTP protocol, and caching, but learning these topics will make any future development much easier.

Finally, Docker is used to deploy our web applications, so basic knowledge of that technology is also useful.

Before we continue, make sure that you have the following applications installed and available on your computer:

- Docker (Community Edition) 17.12.0-ce or later

- Docker Compose (**docker-compose**) 1.18.0 or later

Both these components can be downloaded and installed on all major systems from http://docker.com/. These are essential for completing this activity. Make sure these are available in your system before moving forward.

APPLICATION ARCHITECTURE AND TECHNOLOGIES

In order to deploy our web applications, we will use the tools and technologies described in *Figure 4.4*. Flask is key because it helps us create an HTTP interface for our model, allowing us to access an HTTP endpoint (such as `/predict`) and receive data back in a universal format. The other components are used because they are popular choices when developing web applications:

Tool or Technology	Description	Role
Docker	Docker is a technology used for working with applications packaged in the form of containers. Docker is an increasingly popular technology for building web applications.	Packages Python application and UI.
Flask	Flask is a micro-framework for building web applications in Python.	Creates application routes.
Vue.js	Vue.js is a JavaScript framework that works by dynamically changing templates on the frontend based on data inputs from the backend.	Renders a user interface.
Nginx	Nginx is a web server that is easily configurable to route traffic to Dockerized applications and handle SSL certificates for an HTTPS connection.	Routes traffic between user and Flask application.
Redis	Redis is a key-value database. It's a popular choice for implementing caching systems due to its simplicity and speed.	Cache API requests.

Figure 4.4: Tools and technologies used for deploying a deep learning web application

These components fit together as shown in the following diagram:

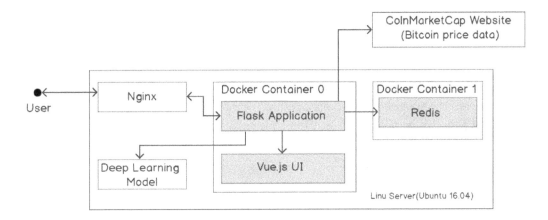

Figure 4.5: System architecture for the web application built in this project

A user visits the web application using their browser. That traffic is then routed by Nginx to the Docker container containing the Flask application (by default, running on port **5000**). The Flask application has instantiated our Bitcoin model at startup. If a model has been given, it uses that model without training; if not, it creates a new model and trains it from scratch using data from Yahoo Finance.

After having a model ready, the application verifies if the request has been cached on Redis; if yes, it returns the cached data. If no cache exists, then it will go ahead and issue predictions, which are rendered in the UI.

EXERCISE 4.02: DEPLOYING AND USING CRYPTONIC

Cryptonic is developed as a dockerized application. In Docker terms, this means that the application can be built as a Docker image and then deployed as a Docker container in either a development or a production environment.

In this exercise, we will see how to use Docker and Cryptonic to deploy the application. Before you begin, download Docker for Desktop from https://www.docker.com/products/docker-desktop Make sure that this application is running in the background before beginning the exercise.

> **NOTE**
> The complete code for this exercise can be found at https://packt.live/2AM5mLP.

1. Docker uses files called **Dockerfiles** to describe the rules for how to build an image and what happens when that image is deployed as a container. Cryptonic's Dockerfile is available in the following code:

> **NOTE**
>
> The triple-quotes (**"""**) shown in the code snippet below are used to denote the start and end points of a multi-line code comment. Comments are added into code to help explain specific bits of logic.

```
FROM python:3.6
ENV TZ=America/New_York
"""

Setting up timezone to EST (New York)
Change this to whichever timezone your data is configured to use.
"""

RUN ln -snf /usr/share/zoneinfo/$TZ /etc/localtime && echo $TZ > /
etc/timezone

COPY . /cryptonic
WORKDIR "/cryptonic"
RUN pip install -r requirements.txt

EXPOSE 5000


CMD ["python", "run.py"]
```

2. A Dockerfile can be used to build a Docker image with the following command:

```
$ docker build --tag cryptonic:latest
```

This command will make the **cryptonic:latest** image available to be deployed as a container. The building process can be repeated on a production server, or the image can be directly deployed and then run as a container.

3. After an image has been built and is available, you can run the Cryptonic application by using the **docker run** command, as shown in the following code:

```
$ docker run --publish 5000:5000 \
--detach cryptonic:latest
```

The **--publish** flag binds port **5000** on localhost to port **5000** on the Docker container, and **--detach** runs the container as a daemon in the background.

In case you have trained a different model and would like to use that instead of training a new model, you can alter the **MODEL_NAME** environment variable on the **docker-compose.yml**. That variable should contain the filename of the model you have trained and want served (for example, **bitcoin_lstm_v1_trained.h5**); it should also be a Keras model. If you do that, make sure to also mount a local directory into the **/models** folder. The directory that you decide to mount must contain your model file.

The Cryptonic application also includes several environment variables that you may find useful when deploying your own model:

MODEL_NAME: Allows us to provide a trained model to be used by the application.

BITCOIN_START_DATE: Determines which day to use as the starting day for the Bitcoin series. Bitcoin prices have a lot more variance in recent years than earlier ones. This parameter filters the data to only years of interest. The default is **January 1, 2017**.

PERIOD_SIZE: Sets the period size in terms of days. The default is **7**.

EPOCHS: Configures the number of epochs that the model trains on every run. The default is **300**.

These variables can be configured in the **docker-compose.yml** file. A part of this file is shown in the following code snippet:

```
version: "3"
   services:
      cache:
         image: cryptonic-cache:latest
         build:
```

```
            context: ./cryptonic-cache
            dockerfile: ./Dockerfile
        volumes:
            - $PWD/cache_data:/data
        networks:
            - cryptonic
    cryptonic:
        image: cryptonic:latest
        build:
            context: .
            dockerfile: ./Dockerfile
        ports:
            - "5000:5000"
        environment:
            - BITCOIN_START_DATE=2019-01-01
            - EPOCH=50
            - PERIOD_SIZE=7
```

4. The easiest way to deploy Cryptonic is to use the **docker-compose.yml** file in the repository (https://packt.live/2AM5mLP).

   This file contains all the specifications necessary for the application to run, including instructions on how to connect with the Redis cache and what environment variables to use. After navigating to the location of the **docker-compose.yml** file, Cryptonic can then be started with the **docker-compose up** command, as shown in the following code:

```
$ docker-compose up -d
```

   The **-d** flag executes the application in the background.

5. After deployment, Cryptonic can be accessed on port **5000** via a web browser. The application has an HTTP API that makes predictions when invoked. The API has the endpoint **/predict**, which returns a JSON object containing the de-normalized Bitcoin price prediction for a week into the future. Here's a snippet showing an example JSON output from the **/predict** endpoint:

```
{
  message: "API for making predictions.",
  period_length: 7,
    result: [ 15847.7,
      15289.36,
```

```
        17879.07,
        …
        17877.23,
        17773.08
    ],
    success: true,
    7
}
```

> **NOTE**
>
> To access the source code for this specific section, please refer
> to https://packt.live/2ZZIZMm.
>
> This section does not currently have an online interactive example, and will
> need to be run locally.

The application can now be deployed on a remote server and you can then use it to continuously predict Bitcoin prices. You'll be deploying an application in the activity that follows.

ACTIVITY 4.01: DEPLOYING A DEEP LEARNING APPLICATION

In this section, based on the concepts explained up to now, try deploying the model as a local web application. You will need to follow these steps:

1. Navigate to the **cryptonic** directory.

2. Build the Docker images for the required components.

3. Change the necessary parameters in **docker-compose.yml**.

4. Deploy the application using Docker on the localhost.

The expected output would be as follows:

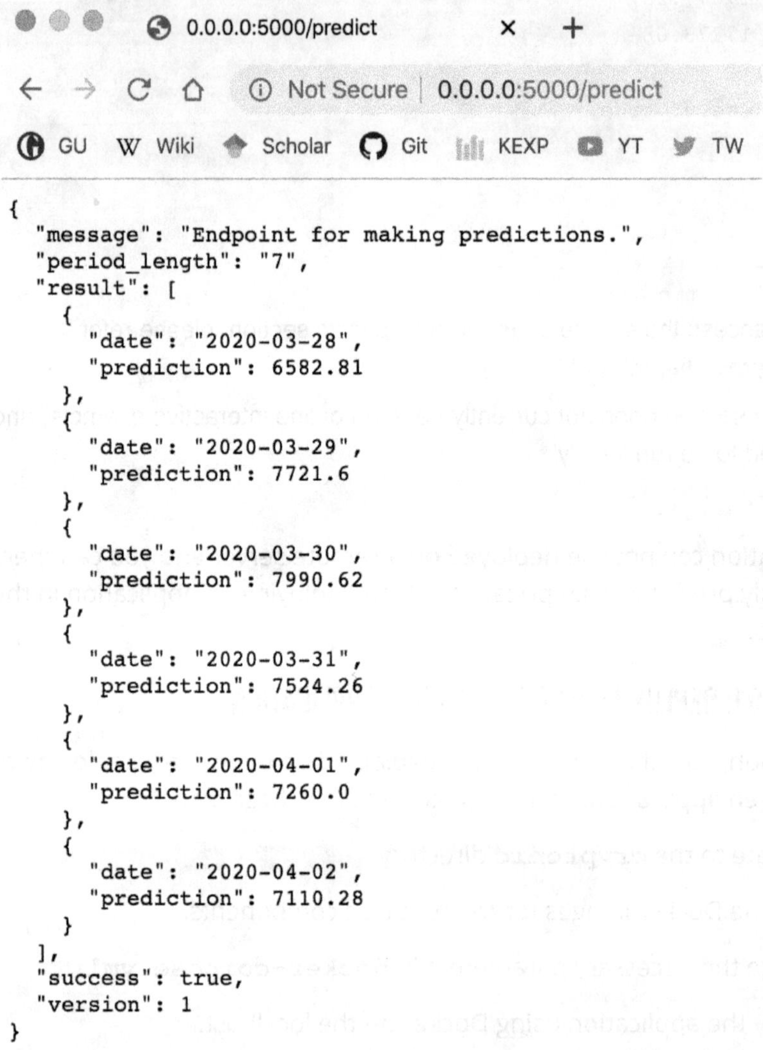
0.0.0.0:5000/predict
Not Secure | 0.0.0.0:5000/predict
GU
Wiki
Scholar
Git
KEXP
YT
TW
{
"message": "Endpoint for making predictions.",
"period_length": "7",
"result": [
{
"date": "2020-03-28",
"prediction": 6582.81
},
{
"date": "2020-03-29",
"prediction": 7721.6
},
{
"date": "2020-03-30",
"prediction": 7990.62
},
{
"date": "2020-03-31",
"prediction": 7524.26
},
{
"date": "2020-04-01",
"prediction": 7260.0
},
{
"date": "2020-04-02",
"prediction": 7110.28
}
],
"success": true,
"version": 1
}

Figure 4.6: Expected output

> **NOTE**
>
> The solution for this activity can be found on page 150.

SUMMARY

This lesson concludes our journey into creating a deep learning model and deploying it as a web application. Our very last steps included deploying a model that predicts Bitcoin prices built using Keras and the TensorFlow engine. We finished our work by packaging the application as a Docker container and deploying it so that others can consume the predictions of our model, as well as other applications, via its API.

Aside from that work, you have also learned that there is much that can be improved. Our Bitcoin model is only an example of what a model can do (particularly LSTMs). The challenge now is twofold: how can you make that model perform better as time passes? And what features can you add to your web application to make your model more accessible? With the concepts you've learned in this book, you will be able to develop models and keep enhancing them to make accurate predictions.

APPENDIX

CHAPTER 01: INTRODUCTION TO NEURAL NETWORKS AND DEEP LEARNING

ACTIVITY 1.01: TRAINING A NEURAL NETWORK WITH DIFFERENT HYPERPARAMETERS

Solution:

1. Using your Terminal, navigate to the directory cloned from https://packt.live/2ZVyf0C and execute the following command to start TensorBoard:

```
$ tensorboard --logdir logs/fit
```

The output is as follows:

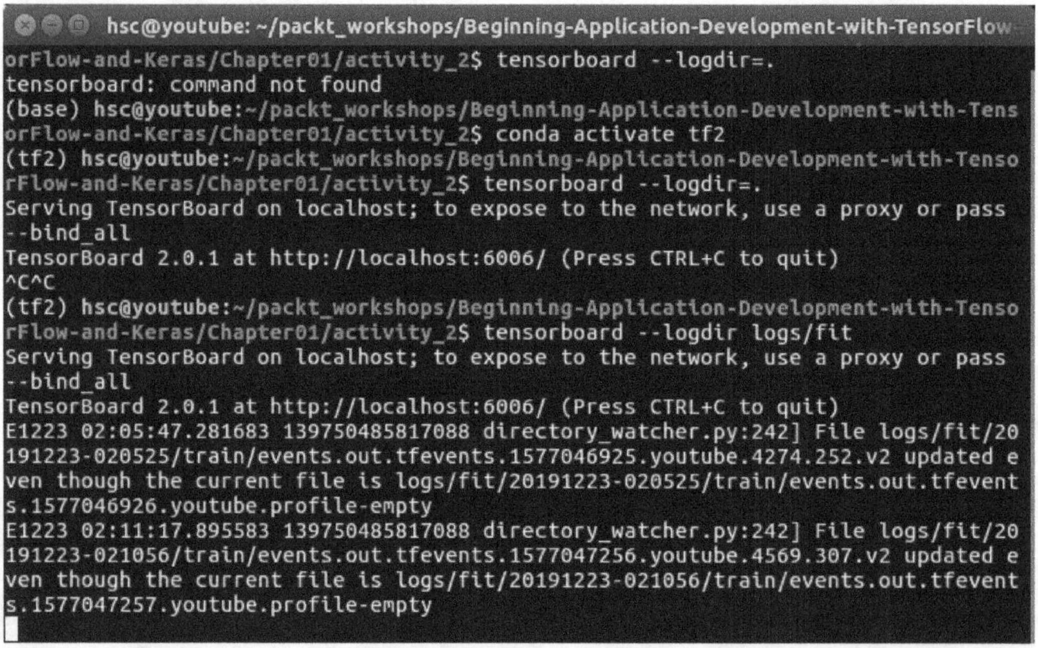

Figure 1.15: A screenshot of a Terminal after starting a TensorBoard instance

2. Now, open the URL provided by TensorBoard in your browser. You should be able to see the TensorBoard **SCALARS** page:

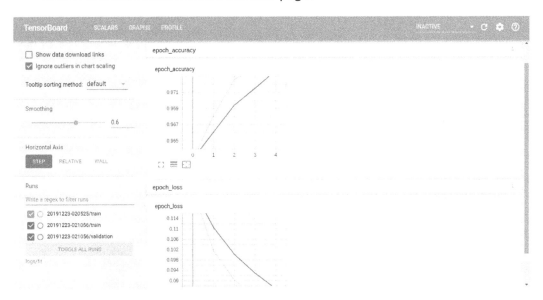

Figure 1.16: A screenshot of the TensorBoard SCALARS page

3. On the TensorBoard page, click on the **SCALARS** page and enlarge the **epoch_accuracy** graph. Now, move the smoothing slider to **0.6**.

The accuracy graph measures how accurately the network was able to guess the labels of a test set. At first, the network guesses those labels completely incorrectly. This happens because we have initialized the weights and biases of our network with random values, so its first attempts are a guess. The network will then change the weights and biases of its layers on a second run; the network will continue to invest in the nodes that give positive results by altering their weights and biases and will penalize those that don't by gradually reducing their impact on the network (eventually reaching **0**). As you can see, this is an efficient technique that quickly yields great results.

4. To train another model by changing various hyperparameters, open a Terminal in **Chapter01/Activity1.01**. Activate the environment. Change the following lines in **mnist.py**:

```
learning_rate = 0.0001 (at line number 47)
epochs = 10 (at line number 56)
```

The **mnist.py** file will look as follows:

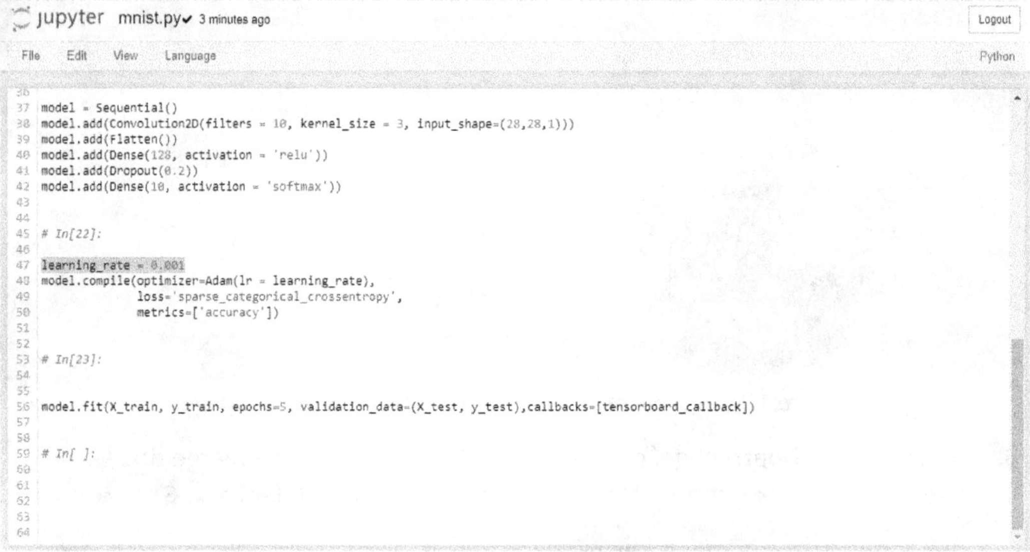

Figure 1.17: A screenshot of the mnist.py file and the hyperparameters to change

5. Now repeat *steps 1-3* for this newly trained model. Start TensorFlow, open the **Scalar** page with the URL seen on TensorBoard, and view the **epoch_accuracy** graph on the **Scalar** page. You will see the difference compared to the earlier graphs:

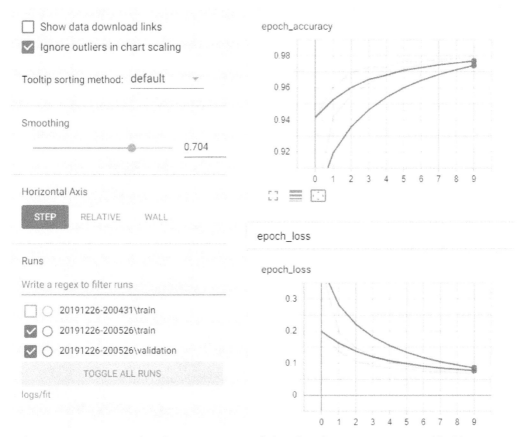

Figure 1.18: A screenshot from TensorBoard showing the parameters specified in step 4

6. Now repeat *step 4*. Open a Terminal in **Chapter01/Activity1.01**. Activate the environment. Change the following lines in **mnist.py** to the following values:

```
learning_rate = 0.01 (at line number 47)
epochs = 5 (at line number 56)
```

Visualize the results. You will get graphs like these:

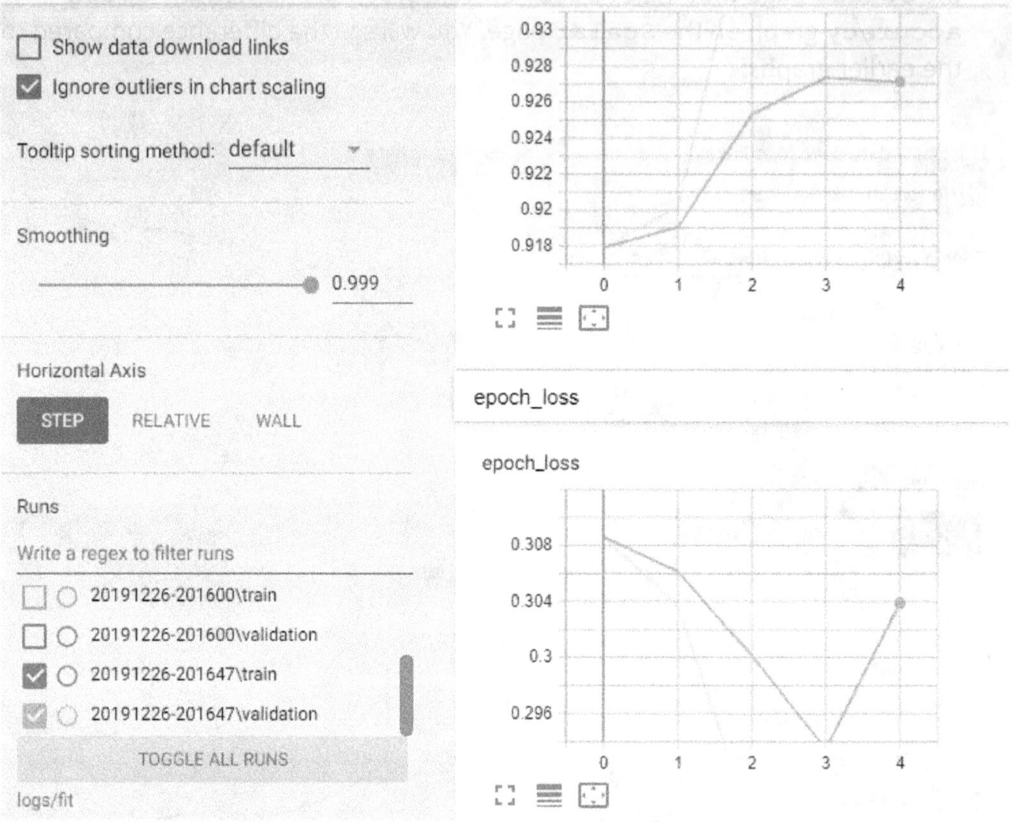

Figure 1.19: A screenshot of the TensorBoard graphs

Now try running the model with any of your custom values and see how the graph changes.

> **NOTE**
>
> Use the `mnist.py` file for your reference at https://packt.live/2ZVyf0C.
>
> There are many other parameters that you can modify in your neural network. For now, experiment with the epochs and the learning rate of your network. You will notice that those two on their own can greatly change the output of your network—but only by so much. Experiment to see if you can train this network faster with the current architecture just by altering those two parameters.

> **NOTE**
>
> To access the source code for this specific section, please refer to https://packt.live/3eiFdC3.
>
> This section does not currently have an online interactive example, and will need to be run locally.

Verify how your network is training using TensorBoard. Alter those parameters a few more times by multiplying the starting values by 10 until you notice that the network is improving. This process of tuning the network and finding improved accuracy is essentially what is used in industry applications today to improve existing neural network models.

CHAPTER 02: REAL-WORLD DEEP LEARNING: PREDICTING THE PRICE OF BITCOIN

ACTIVITY 2.01: ASSEMBLING A DEEP LEARNING SYSTEM

Solution:

We will continue to use Jupyter Notebooks and the data prepared in previous exercises of chapter 2 (**data/train_dataset.csv**), as well as the model that we stored locally (**bitcoin_ lstm_v0.h5**):

1. Import the libraries required to load and train the deep learning model:

```
import numpy as np
import pandas as pd
import matplotlib.pyplot as plt
%matplotlib inline
from tensorflow.keras.models import load_model
plt.style.use('seaborn-white')
```

> **NOTE**
>
> The **close_point_relative_normalization** variable will be used to train our LSTM model.

We will start by loading the dataset we prepared during our previous activities. We'll use pandas to load that dataset into memory.

2. Load the training dataset into memory using pandas, as follows:

```
train = pd.read_csv('data/train_dataset.csv')
```

3. Now, quickly inspect the dataset by executing the following command:

```
train.head()
```

As explained in this chapter, LSTM networks require tensors with three dimensions. These dimensions are period length, the number of periods, and the number of features.

Now, proceed to create weekly groups, and then rearrange the resulting array to match those dimensions.

4. Feel free to use the **create_groups()** function provided to perform this operation:

create_groups(data=train, group_size=7)

The default values for that function are **7** days.

Now, make sure you split the data into two sets: training and validation. We do this by assigning the last week from the Bitcoin prices dataset to the evaluation set. We then train the network to evaluate that last week. Separate the last week of the training data and reshape it using **numpy.reshape()**. Reshaping it is important, as the LSTM model only accepts data organized in this way:

```
X_train = data[:-1,:].reshape(1, 186, 7)
Y_validation = data[-1].reshape(1, 7)
```

Our data is now ready to be used in training. Now, we load our previously saved model and train it with a given number of epochs.

5. Navigate to the **Load Our Model** header and load our previously trained model:

```
model = load_model('bitcoin_lstm_v0.h5')
```

6. And now, train that model with our training data, **X_train** and **Y_validation**:

```
%%time
history = model.fit( x=X_train, y=Y_validation, epochs=100)
```

Notice that we store the logs of the model in a variable called **history**. The model logs are useful for exploring specific variations in its training accuracy and observing how well the loss function is performing:

Make Predictions

```
[21]:  %%time
       history = model.fit(
           x=X_train, y=Y_validation,
           epochs=100)

       Train on 1 samples
       Epoch 1/100
       1/1 [==============================] - 0s 241ms/sample - loss: 0.0096
       Epoch 2/100
       1/1 [==============================] - 0s 25ms/sample - loss: 0.0085
       Epoch 3/100
       1/1 [==============================] - 0s 28ms/sample - loss: 0.0078
       Epoch 4/100
       1/1 [==============================] - 0s 33ms/sample - loss: 0.0072
       Epoch 5/100
       1/1 [==============================] - 0s 32ms/sample - loss: 0.0067
       Epoch 6/100
       1/1 [==============================] - 0s 31ms/sample - loss: 0.0062
       Epoch 7/100
       1/1 [==============================] - 0s 32ms/sample - loss: 0.0058
       Epoch 8/100
       1/1 [==============================] - 0s 33ms/sample - loss: 0.0054
       Epoch 9/100
       1/1 [==============================] - 0s 32ms/sample - loss: 0.0051
       Epoch 10/100
       1/1 [==============================] - 0s 33ms/sample - loss: 0.0047
       Epoch 11/100
       1/1 [==============================] - 0s 42ms/sample - loss: 0.0044
       Epoch 12/100
       1/1 [==============================] - 0s 33ms/sample - loss: 0.0041
```

Figure 2.27: Section of the Jupyter Notebook where we load our earlier model and train it with new data

7. Finally, let's make a prediction with our trained model. Using the same data, **X_train**, call the following method:

```
model.predict(x=X_train)
```

8. The model immediately returns a list of normalized values with the prediction for the next 7 days. Use the **denormalize()** function to turn the data into US dollar values. Use the latest values available as a reference for scaling the predicted results:

```
denormalized_prediction = denormalize(predictions, \
                                      last_weeks_value)
```

The output is as follows:

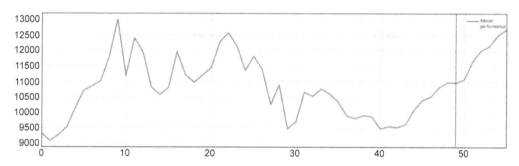

Figure 2.28: Projection of Bitcoin prices for 7 days in the future
using the LSTM model we just built

> **NOTE**
>
> We combine both time series in this graph: the real data (before the vertical line) and the predicted data (after the vertical line). The model shows a variance similar to the patterns seen before and it suggests a price increase during the following 7-day period.

9. When you are done experimenting, save your model with the following command:

```
model.save('bitcoin_lstm_v0_trained.h5')
```

We will save this trained network for future reference and compare its performance with other models.

The network may have learned patterns from our data, but how can it do that with such a simple architecture and so little data? LSTMs are powerful tools for learning patterns from data. However, we will learn in our next sessions that they can also suffer from overfitting, a phenomenon common in neural networks, in which they learn patterns from the training data that are useless when predicting real-world patterns. We will learn how to deal with that and how to improve our network to make useful predictions.

> **NOTE**
>
> To access the source code for this specific section, please refer to https://packt.live/2ZWfqub.
>
> You can also run this example online at https://packt.live/3glhhcT.
> You must execute the entire Notebook in order to get the desired result.

CHAPTER 3: REAL-WORLD DEEP LEARNING: EVALUATING THE BITCOIN MODEL

ACTIVITY 3.01: OPTIMIZING A DEEP LEARNING MODEL

Solution:

1. Using your Terminal, start a TensorBoard instance by executing the following command:

```
$ cd ./Chapter03/Activity3.01/
$ tensorboard --logdir=logs/
```

You will see the **SCALARS** page once TensorBoard opens in the browser:

Figure 3.20: Screenshot of a TensorBoard showing SCALARS page

2. Open the URL that appears on screen and leave that browser tab open as well. Also, start a Jupyter Notebook instance with the following command:

```
$ jupyter-lab
```

Here's the screenshot showing the Jupyter Notebook:

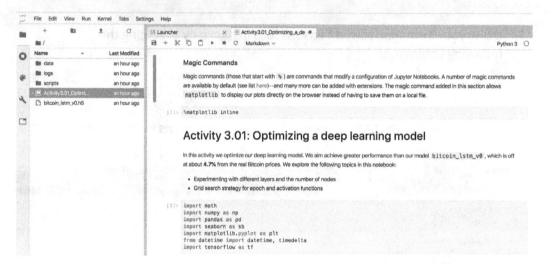

Figure 3.21: Jupyter Notebook

Open the URL that appears in a different browser window.

3. Now, open the Jupyter Notebook called **Activity3.01_Optimizing_a_ deep_learning_model.ipynb** and navigate to the title of the Notebook. Run the cell to, import all the required libraries.

4. Set the seed to avoid randomness:

```
np.random.seed(0)
```

We will load the train and test data like we did in the previous activities. We will also split it into train and test groups using the **split_lstm_input()** utility function:

Load Data

We will load our same train and testing set from previous activitites.

```
In [7]:  train = pd.read_csv('data/train_dataset.csv')

In [8]:  test = pd.read_csv('data/test_dataset.csv')

In [9]:  train_data = create_groups(
             train['close_point_relative_normalization'].values, 7)

In [10]: test_data = create_groups(
             test['close_point_relative_normalization'].values, 7)

In [11]: X_train, Y_train = split_lstm_input(train_data)

In [12]: X_train.shape, Y_train.shape
Out[12]: ((1, 186, 7), (1, 7))
```

Figure 3.22: Screenshot showing results of loading datasets

In each section of this Notebook, we will implement new optimization techniques in our model. Each time we do so, we'll train a fresh model and store its trained instance in a variable that describes the model version. For instance, our first model, **bitcoin_lstm_v0**, is called **model_v0** in this Notebook. At the very end of the Notebook, we'll evaluate all the models using MSE, RMSE, and MAPE.

5. To get these models up and running, execute the cells under the **Reference Model** section.

6. Now, in the open Jupyter Notebook, navigate to the **Adding Layers and Nodes** header. You will recognize our first model in the next cell. This is the basic LSTM network that we built in *Chapter 2, Real-World Deep Learning: Predicting the Price of Bitcoin*. Now, we have to add a new LSTM layer to this network:

```
[17]:  period_length = 7
       number_of_periods = 186
       batch_size=1
```

```
[18]:  model_v1 = Sequential()
       model_v1.add(LSTM(
           units=period_length,
           batch_input_shape=(batch_size, number_of_periods, period_length),
           input_shape=(number_of_periods, period_length),
           return_sequences=True, stateful=False))

       #
       #  Add new LSTM layer to this network here.
       #
       model_v1.add(LSTM(
           units=period_length,
           batch_input_shape=(batch_size, number_of_periods, period_length),
           input_shape=(number_of_periods, period_length), stateful=False))

       model_v1.add(Dense(units=period_length))
       model_v1.add(Activation("linear"))

       model_v1.compile(loss="mse", optimizer="rmsprop")
```

```
[22]:  %%time
       train_model(model=model_v1, X=X_train, Y=Y_train, epochs=100, version=1, run_number=0)

       CPU times: user 18.1 s, sys: 3.06 s, total: 21.2 s
       Wall time: 11.4 s
[22]:  <tensorflow.python.keras.callbacks.History at 0x15d28b610>
```

Figure 3.23: Jupyter Notebook with code for adding new LSTM layer

Using your knowledge from this chapter, go ahead and add a new LSTM layer and then compile and train the model.

While training your models, remember to frequently visit the running TensorBoard instance. You will be able to see each model run and compare the results of their loss functions there:

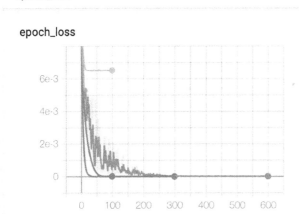

Figure 3.24: Output of the loss function for different models

The TensorBoard instance displays many different model runs. TensorBoard is really useful for tracking model training in real time.

7. In this section, we are interested in exploring different magnitudes of epochs. Use the **train_model()** utility function to name different model versions and runs:

```
train_model(model=model_v0, X=X_train, \
            Y=Y_validate, epochs=100, \
            version=0, run_number=0)
```

Train the model with a few different epoch parameters.

At this point, you are interested in making sure the model doesn't overfit the training data. You want to avoid this, because if it does, it will not be able to predict patterns that are represented in the training data but have different representations in the test data.

After you are done experimenting with epochs, move to the next optimization technique: activation functions.

8. Now, navigate to the **Activation Functions** header in the Notebook. In this section, you only need to change the following variable:

```
activation_function = "relu"
```

We have used the **tanh** function in this section, but feel free to try other activation functions. Review the list available at https://keras.io/activations/ and try other possibilities.

Our final option is to try different regularization strategies. This is notably more complex and may take a few iterations for us to notice any gains—especially with so little data. Also, adding regularization strategies typically increases the training time of your network.

9. Now, navigate to the **Regularization Strategies** header in the Notebook. In this section, you need to implement the **Dropout()** regularization strategy. Find the right place to put that step and implement it in our model:

Regularization Strategies

In this section we implement a `Dropout()` regularization strategy.

```
In [24]:  model_v4 = Sequential()
          model_v4.add(LSTM(
              units=period_length,
              batch_input_shape=(batch_size, number_of_periods, period_length),
              input_shape=(number_of_periods, period_length),
              return_sequences=True, stateful=False))

          #
          #  Implement a Dropout() here.
          #
          model_v4.add(Dropout(0.4))

          model_v4.add(LSTM(
              units=period_length,
              batch_input_shape=(batch_size, number_of_periods, period_length),
              input_shape=(number_of_periods, period_length),
              return_sequences=False, stateful=False))

          #
          #  Implement a Dropout() here too.
          #
          model_v4.add(Dropout(0.4))
```

Figure 3.25: Jupyter Notebook showing code for regularization strategies

10. You can also try L2 regularization here (or combine both).
    Do the same as you did with **Dropout()**, but now using
    ActivityRegularization(l2=0.0001).

    Finally, let's evaluate our models using RMSE and MAPE.

11. Now, navigate to the **Evaluate Models** header in the Notebook. In this
    section, we will evaluate the model predictions for the next 19 weeks of data in
    the test set. Then, we will compute the RMSE and MAPE of the predicted series
    versus the test series.

    First plot looks as follows:

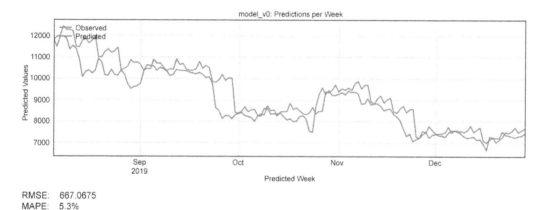

RMSE: 667.0675
MAPE: 5.3%

Figure 3.26: Prediction series versus test series #1

Second plot looks as follows:

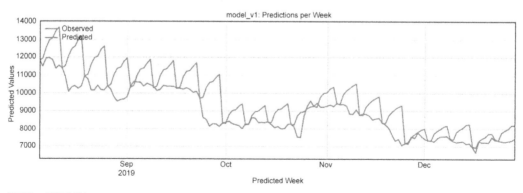

RMSE: 1123.3164
MAPE: 9.3%

Figure 3.27: Prediction series versus test series #2

Third plot looks as follows:

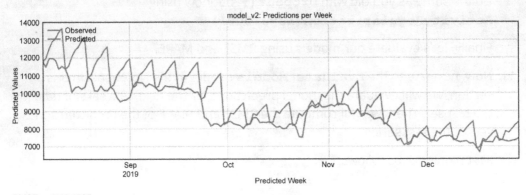

RMSE: 1116.1967
MAPE: 9.2%

Figure 3.28: Prediction series versus test series #3

Fourth plot looks as follows:

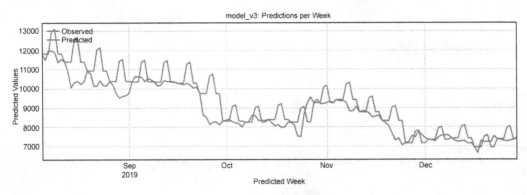

RMSE: 792.1771
MAPE: 6.0%

Figure 3.29: Prediction series versus test series #4

Fifth plot will look as follows:

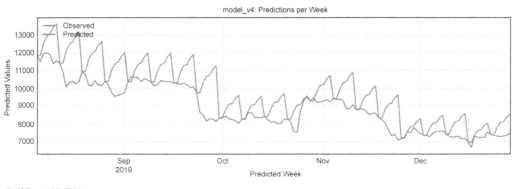

RMSE: 1188.5708
MAPE: 10.0%

Figure 3.30: Prediction series versus test series #5

We have implemented the same evaluation techniques from *Exercise 2.01, Exploring Bitcoin Dataset*, https://packt.live/3ehbgCi, all wrapped in utility functions. Simply run all the cells from this section until the end of the Notebook to see the results.

> **NOTE**
>
> To access the source code for this specific section, please refer to https://packt.live/2ZgAo87.
>
> You can also run this example online at https://packt.live/3ft5Wgk.
> You must execute the entire Notebook in order to get the desired result.

In this activity, we used different evaluation techniques to get more accurate results. We tried to train for more epochs, changed the activation function, added regularization, and compared results in different scenarios. Take this opportunity to tweak the values for the preceding optimization techniques and attempt to beat the performance of that model.

CHAPTER 4: PRODUCTIZATION

ACTIVITY 4.01: DEPLOYING A DEEP LEARNING APPLICATION

Solution:

In this activity, we deploy our model as a web application locally. This allows us to connect to the web application using a browser or to use another application through the application's HTTP API. You can find the code for this activity at https://packt.live/2Zdor2S.

1. Using your Terminal, navigate to the **cryptonic** directory and build the Docker images for all the required components:

```
$ docker build --tag cryptonic:latest .
$ docker build --tag cryptonic-cache:latest        cryptonic-cache/
```

Those two commands build the two images that we will use in this application: **cryptonic** (containing the Flask application) and **cryptonic-cache** (containing the Redis cache).

2. After building the images, identify the **docker-compose.yml** file and open it in a text editor. Change the **BITCOIN_START_DATE** parameter to a date other than **2017-01-01**:

```
BITCOIN_START_DATE = # Use other date here
```

3. As a final step, deploy your web application locally using **docker-compose up**, as follows:

```
docker-compose up
```

You should see a log of activity on your Terminal, including the training epochs completed by your model.

4. After the model has been trained, you can visit your application at **http:// localhost:5000** and make predictions at **http://localhost:5000/predict**:

● ● ●   🌐 0.0.0.0:5000/predict   ✕   +

← → C ⌂   ⓘ Not Secure | 0.0.0.0:5000/predict

🅖 GU   W Wiki   ◆ Scholar   ⓞ Git   ▮▮ KEXP   ▶ YT   🐦 TW

```
{
  "message": "Endpoint for making predictions.",
  "period_length": "7",
  "result": [
    {
      "date": "2020-03-28",
      "prediction": 6582.81
    },
    {
      "date": "2020-03-29",
      "prediction": 7721.6
    },
    {
      "date": "2020-03-30",
      "prediction": 7990.62
    },
    {
      "date": "2020-03-31",
      "prediction": 7524.26
    },
    {
      "date": "2020-04-01",
      "prediction": 7260.0
    },
    {
      "date": "2020-04-02",
      "prediction": 7110.28
    }
  ],
  "success": true,
  "version": 1
}
```

Figure 4.7: Screenshot of the Cryptonic application deployed locally

NOTE

To access the source code for this specific section, please refer to https://packt.live/2Zg0wjd.

This section does not currently have an online interactive example, and will need to be run locally.

INDEX